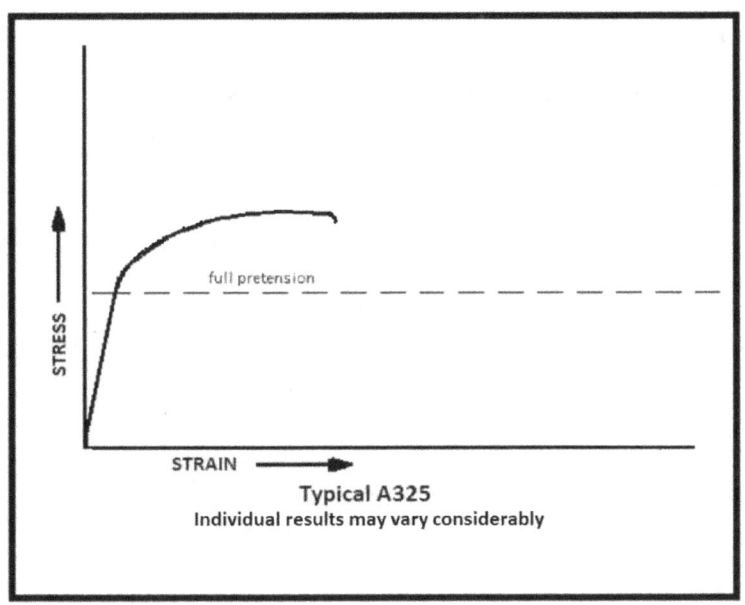

A SHORT GUIDE TO HIGH STRENGTH BOLTING

by

Ethan Moore

A SHORT GUIDE TO HIGH STRENGTH BOLTING

by Ethan Moore

Copyright 2018

Equation 5.4 of the RCSC Specification used with permission

To all those inspectors, contractors, *and the occasional engineer,* who make life so interesting and instructions so pointless...

Table of Contents

WHAT ARE WE TALKING ABOUT HERE? 6

THE STRESS-STRAIN CURVE: 16

BOLTED JOINTS 22

FORCES ON BOLTS 29

PRETENSIONING: where the fun begins. 39

COMMENTS ON SNUGTIGHT JOINTS 53

SOME COMMENTS ON "SLIP-CRITICAL" 59

IS IT RIGHT? 65

    PRETENSIONED JOINTS 65

    Pretensioned Joints: effect oriented methods 72

    PRETENSIONED JOINTS: TORQUE ORIENTED METHODS 80

PRE-INSTALLATION VERIFICATION 90

    TURN-OF-NUT 96

    F1852/ F2280 METHOD 102

    ONE FINAL METHOD: DTI WASHERS 104

NO FURTHER VERIFICATION? WHEN? 106

    NO FURTHER VERIFICATION REQUIRED? What about prior verification? 114

    Turn-Of-Nut: periodic or continuous inspection? 118

    Direct Tension Indicator Washers: periodic or continuous inspection? 119

Torque Wrench – "Calibrated" Wrench method: periodic or continuous inspection? 121

F1852 and F2280 TC Bolts: periodic or continuous inspection? 121

Tolerances? 125

Concluding Remarks 140

FURTHER READING 142

WHAT ARE WE TALKING ABOUT HERE?

Could there be anything more depressing than High Strength Bolting? There is a feeling of failure before it even begins. It is perhaps the least understood matter in steel construction. And, when something is not understood, either by contractors or inspectors, it often gets ignored completely. Yet, the information is available to all involved if they would look at it. All they have to do to understand it is do their jobs with some faith in the engineers and scientists who performed the research and experiments which the principles of high strength bolting are founded upon, and understanding what they are doing is part of their jobs. The information is available in various forms, but the best forms are meant more for engineers than for contractors or inspectors. Of those, ***it is obvious from many studies and the experience of the experts that any method which places total reliance on torque as a method of installation or verification is grossly inadequate and misleading and unreliable.*** Some writings exist for contractors, but they are inadequate and are grossly misunderstood. They are written only as "how to" and not intended to help the contractors understand what they are doing and the very important reasons for doing things a certain way. But, to do their jobs right, and that includes inspectors, understanding on the deeper level should not be necessary, in theory, since they should be following directions. To do their jobs right they only have to do what the codes and specifications say and not do what they say not to do. But, what people don't understand is usually considered unimportant. People usually won't just trust one's word that these methods are critical to the performance of a bolted joint, especially if some things seem counter-intuitive when not considered in the proper light. It is more difficult to get the point across in high strength bolting than in other construction methods because the point of the rules is to create an effect which is invisible. It is easy to understand something visible like welding standards when one can see the cracks or porosity on the surface telling you that something is wrong which no one can *honestly* deny. Welding is considered important and is taken more seriously by contractors and inspectors than bolting. Part of it could be because of the strict certification requirements for welding, but no contractors are required to certify for bolting to prove that they understand what they are doing. In bolting, you can't see the pressure or slip resistance of a joint just by looking at it, unless it is so bad that the plates are not in contact. Any number of false assumptions of the adequacy of what an ironworker is doing or not doing will be stated as a reason why none of the rules matter. No document will ever make them see the light unless it is read in the proper perspective. This failure in steel construction is why I am writing this. People are not just not on the same page, they are not even in the same book – they are in some book I have never seen.

The most common dangerously false assumptions are: 1) a looser bolt is safe and less likely to break than a bolt pretensioned to the *required minimum*, as if the contractor knows more than the scientists and engineers who developed the codes, 2) inspection of all stages is not required since there is the occasional word "periodic" stated under only very specific circumstances and even then at only particular stages of the process, and therefore there is little to no cooperation with inspectors, 3) that every method is *prequalified* and that one is allowed to proceed without the Mandatory Pre-Installation Verification testing, 4) that it is the *bolts* being qualified and not the *procedure*, 5) that one can use charts to pretension bolts without verification testing of any kind or that one can use a chart to disregard the procedure qualified if one doesn't like reality, even though such use of charts is not allowed, 6) that F1852 bolts take care of themselves and have fewer rules or none at all when they actually have more rules and are less predictable, 7) that one can simply test a few joints after the building is up, when there was no inspection at all, and the whole building would then be passed, when 10% testing applies to every joint, and if tested according to the rules, much more than 10% of the bolts would be tested, even though without inspection beginning from the first day

(Mandatory Pre-Installation Verification) <u>nothing</u> can be passed at all by an inspector and that matter is up to the engineer and building official to resolve since that is a major violation, 8) that proper snugging of all bolts in a joint before pretensioning has no importance.

The effects of following proper procedures are invisible, just as the effects of not following the mandatory procedures. Both contractors and inspectors, when neither understands the reason for the procedures, don't see the difference between a safe joint which will behave as designed and one that won't, until something bad happens and it is too late. Because the effects are invisible, it is even more critical that the procedures be followed and the necessary verifications are performed as required. One can't assume that a joint is safe just by looking at it after it is together with no consideration of how it got there.

Some people may be offended by my writing this and discussing the failures in this industry. There is no nice way to discuss these failures. I am not writing this to make friends. I am writing this to address very serious problems which are being ignored to a great extent. Some may be offended because this implies that they have been doing things wrong, either deliberately or unintentionally, for a very long time, and it calls into question the adequacy of what they have done and are responsible for. If this hits a nerve – good. It implies that people don't understand something simple. There are many people who insist that they are doing everything perfectly even though they have no knowledge of what a bolt does, and those will probably be most offended, since they will see that they have no excuse. You know who you are.

Many people argue that the methods commonly used are just fine and meet every requirement, while ignoring every one of those requirements or grossly misinterpreting the ones they feel like acknowledging. People, including inspectors, often feel that the engineer who designed that structure is wrong by specifying a pretensioning method which is more reliable. They say that the way they always do it must be absolutely right, even when not approved by the engineer.

Many inspectors and instructors of inspectors may be offended because they passed an ICC or the older ICBO test, and that must mean that they know everything already, even if they can't apply it on an actual project. Of course, being certified *should* imply an in-depth knowledge of structural bolting, since the exams are supposed to test a person's knowledge of the subject. In reality, in the past before the ICC made a separate certification for bolting apart from welding, it was easy for anyone to pass the exam without any knowledge of bolting, since that was only a small part of that exam. That matter has been addressed to some extent by splitting the exams, besides the fact that they can now charge two fees. People can still be fed answers in a class and be able to mindlessly spit them out on a test, but on a real project, those are just empty words. Welding is still the subject considered of singular importance, bolting even being considered too simple for consideration. Ink on a wallet card has no meaning if it can't produce better safety and quality of construction. Many areas of the country have no certification requirements for inspectors, or only require that an inspector be a "knowledgeable" individual, whatever that means. It takes the correct perspective to apply any of those rules correctly and produce the effect intended in design. Without the correct perspective, one is living in a fantasy world no matter what answers they were told by a worthless irresponsible instructor to spit out during a test. One should be attempting to understand the principles of bolting as the engineer understands them before questioning any of the principles used in design or specified in the procedure. Unless they are the engineer, or the scientist performing the experiments, they don't have any place deciding to throw out or modify requirements.

It is very common – actually universal – for people to think that torque or rotation is the point of the procedure. There is particular emphasis on torque among contractors, which is least reliable and is often tied to charts which have been declared unusable, and are even labeled so. When one looks at the RCSC Specification for Structural Joints Using ASTM A325 or A490 Bolts (hereafter noted as RCSC Specification), which is required to be followed by code, one sees that it is **tension** and the **pressure** it creates which the experts focus on. One does not need to be an engineer to see that. There are no formulas which use torque or rotation in the *design* of joints.

The most serious and criminal failure is the belief that a looser bolt is safer than a tighter one. That is beyond stupid. It is very common – actually nearly universal – for contractors to *refuse* to meet what they *know* is the Code Required Minimum, saying that it will weaken the bolt break it and that it is unsafe. That implies, among other things, that they know nothing about these bolts we are talking about or about how these bolts behave under load. That is inexcusable, especially since they know that they are not meeting the minimum requirements. Why do the codes exist if not following them is better in an earthquake? People forget that the codes were written in blood. Many people pretending to be professional won't even study these matters because they believe bolting to be too simple to care about. Do they know more than the scientists and engineers who studied the failures of the past and are trying to prevent them from happening again? How cheap is a building or equipment which is unusable because of shifts of the joints? It is inexcusable for people to play such games.

Bolting is quite simple, if one would seek to understand it properly. There are fewer rules than for something such as welding, and joint design can be much simpler and more ductile. One does not have to understand it as long as one intends to follow all requirements without argument, trusting that there are important reasons for everything.

Perhaps a serious reason why true knowledge and understanding of the rules and procedures is so rare among contractors is because it is rare among inspectors. If the inspectors did their jobs, the contractors would do their jobs whether they understood it or not.

In the simplest terms, inspectors are there to see that the engineer's intentions, as approved by the building official or the jurisdiction in authority, are followed. It is an unfortunate fact that many companies performing this service are on a job only to get paid, and not because they care about the result. In a certain sense, it has to be understood that no one is going to do it for free and everyone needs money to take care of themselves and those they care about, and that includes the ones who designed the structure. But, it can be easy and extremely common to go too far into ethical violations. The inspectors had made an oath to place the public welfare above all else. Fraud in inspection is a common way for companies to try to save money and make a profit in direct and conscious violation of that oath, and this is much worse for an activity which is not viewed as critical, but this is not the place to go into a detailed discussion of ethical matters, although those matters have a large impact on both the safety and performance of joints, and played at least an indirect role in the need to write this. Ethical matters also have an influence on individuals recognizing the need to fully understand the principles of high strength bolting. Fraud can't be excused just because someone made a bidding error. In this kind of situation discussed in this paragraph there is a great danger when a company does not care about the results and only cares about the numbers on an invoice, an invoice which is supposed to represent a performance of duty. They can entirely forget the whole reason they are being paid and the purpose behind hiring such companies for inspection and testing.

No method of bolting is reliable to any degree in and of itself. Reliability of any method depends entirely on the inspector performing every required verification at every stage of the process with complete cooperation from the contractor. It doesn't matter how well the engineer performs the design calculations when the one who is the eyes and ears of the engineer on the jobsite (the inspector) won't do anything to ensure that the engineer's intentions are carried out. If they consistently looked to applying the engineer's intentions, then contractors would not be so shocked when a REAL inspector shows up and expects them to follow the rules and give complete cooperation for all the **mandatory** requirements. That doesn't happen often enough, and the fault often lies with the inspectors.

The apparent invisibility of the reliability of a joint after the work has been done – a matter to be discussed in more detail in later sections – only makes it easier to say that it was done right after the fact. Inspectors who failed to inspect very often cave in under the pressure of their failure and are commonly inclined to simple sign it off after the fact, not because they observed anything being done right, but because they don't know that the procedures were not followed, which is itself a violation that they are reluctant to admit. There is pressure from the laboratories involved to have the inspector only show up after all the verification has been skipped in order to save money, and demand that the inspector sign it under threat of being fired. This is too great of a temptation for many when money is involved, since in the end, a signature on a piece of paper is the only thing people, including the engineer, look at when the job is over, since the engineer is usually unable to witness the work at every necessary stage personally – hence the requirement for special inspectors. Everyone knows that, in the end, the engineers and building officials will just take their word for it without question, since to do otherwise without sufficient evidence that the duties of the inspectors and contractors were not performed as required would involve very costly delays and extra testing which someone has to pay for. I expect many people to be offended by my writing the plain truth about what is really going on, and I expect that they would be very defensive over this issue. Those who I am talking about know who they are. The best defense for them would be to make sure they do their jobs right if they don't want to feel guilty.

This is about what happens and about what is supposed to happen when high strength bolts are used in a structure.

I intend to fill the gap between engineered design and what occurs in reality. I do not intend this to be a manual for design of bolted joints, and there are many great design manuals out there already. I also do not intend this to be a "how to" type of work. My desire is to broaden the understanding of the principles of bolting for those involved in the field so that they can have greater understanding of the requirements, and therefore be better able to meet those requirements. I will try to keep math to a minimum.

I will be focusing more attention on seismic situations than on other situations, but that is not the only case when high strength bolting is used or when the bolts are pretensioned. Bridges also come to mind, and there are other situations that don't have to include what people think of as buildings. The seismic applications are a situation where the failures to understand and follow the required procedures are quite dangerous and present a threat which is more than just a usability issue. The requirements of high strength bolting, pretensioned of not, are there to transfer a design load from one member to another, and the joint must therefore be capable of functioning as designed. The load may be from wind, vibration, gravity, earthquake or other situation, but if a load of any magnitude exists, the joint must be capable of handling it and transferring it in the manner that the engineer assumed it would, whether or not the design requirements are understood by contractors and inspectors. "Snugtight" and "slip-critical" can be the difference between failure through the smaller net section of a plate or failure through the larger gross section as though the holes weren't there.

I am writing this from the assumption that inspection of bolting is required or specified for a particular bolting project. Even if the contractor is being allowed to erect the structure alone without third party verification, all of this still applies. Bolting is bolting. It must still be understood and done right. The principles don't change from one project to another, or from one jurisdiction to another. The laws of physics don't change, no matter what a human being writes.

I will not be discussing such things as strain gages or ultrasonic measurement, more advanced tension or strain controlled bolt assemblies or some of the other technological advances which can make this much simpler – for a price. I am confining this discussion to the methods and bolts most commonly used in the United States. It is those methods which make this necessary.

Be warned, in this discussion repetition is necessary.

There is sometimes concern over the continuing appropriateness of using a single gender specific pronoun to imply everyone of any gender. I will avoid the issue by using the neutral pronoun   "it."

THE STRESS-STRAIN CURVE:

The Foundation of Everything

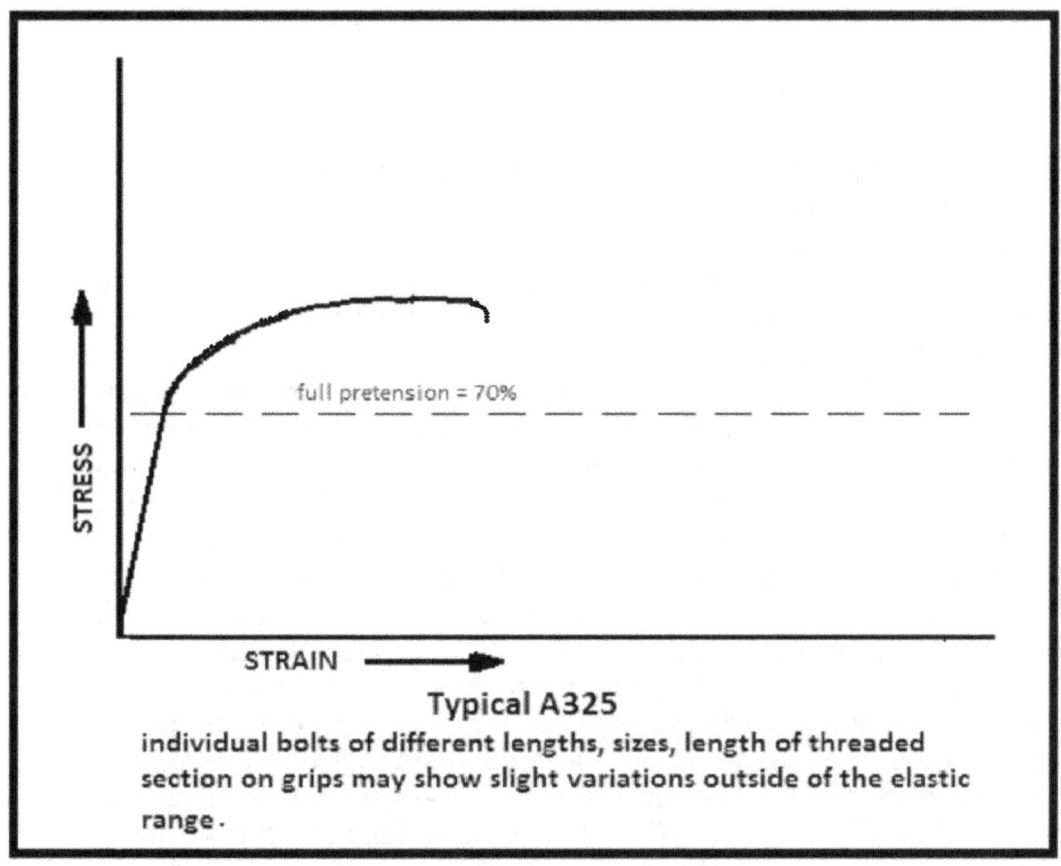

Typical A325
individual bolts of different lengths, sizes, length of threaded section on grips may show slight variations outside of the elastic range.

Since this has to do with materials, let's look at some.

1. 3 Basic Types:
   A325, A490, F1852/F2280

F1852 and F2280 are not entirely a different type, but are defined as an application of the other standards and must meet the requirements of the others. A449 almost doesn't count, but has common qualities of the others, but is not made to the same standards or to the same heavy hex dimensions as the others, and therefore has restrictions on use. A449 is used when specified diameter is over 1.5", which is the maximum size of the other standards, or in some cases involving ridiculously long bolts such as with some tube connections. Of course, nuts and washers must also meet strength requirements, since the "assembly" is only as strong as the weakest part. A nut must be able to take the load without stripping. F1852 bolts come as an assembly, assembled by the manufacturer and each piece must be used as a single unit, and no mixing of parts from other lots is allowed since the manufacturer only guarantees the product as assembled.

2. Material Properties:

High strength bolts are made from medium carbon quenched and tempered steel, with other alloying elements, and can be plain, coated (with limitations), or alloyed for greater corrosion resistance (type 3). These bolts have a much greater strength than ordinary A307 bolts, and also a much greater strength than the usual grade 50 steel being connected.

The strengths of these bolts are affected by diameter. As the steel is being quenched, and cooling after tempering, thinner sections of steel cool faster, thereby retaining a greater hardness than thicker bolts that cool slower. Therefore, for A325 bolts above 1" diameter, softer steel is expected and the specified strength is reduced. A325 has minimum tensile strength of 120ksi for diameters up to and including 1", and 105ksi for diameters over 1". A490 bolts have even greater strength at 150ksi for all diameters. It is significant that the dividing line for the strength of A325 bolts was drawn at 1" and that those were included in the stronger group, because they only barely meet the 120ksi standard and represent the limit of possibility, particularly with certain manufacturers. The result is that if any F1852 TC bolt is going to fail testing it will most likely be the 1" bolts. That is something to always keep in mind.

Threads also affect the behavior of the bolts, which can even be observed to some degree on a Skidmore test. Each diameter of bolt has a particular thread pitch. Each bolt therefore has a particular ratio of thread pitch to diameter. When comparing a ½ inch A325 to a 1 inch A325 the number of threads per diameter of length increases from 6.5 to 8, which is a significant difference. That means that when speaking in terms of diameters of length, which is referred to in the AISC and RCSC Specifications, for a given degree of rotation there is a greater strain per diameter than with the larger bolts, all other factors being equal. That is noticeable when comparing bolts of significant difference in diameter but not noticeable when comparing a 7/8 inch A325 with a 1 inch A325. We are here comparing the thread pitch to diameter *ratio* rather than the thread pitch to thread pitch, the size of which of course increases as the diameter increases and are not to be compared directly.

3. Stress-Strain Curve: the key to understanding bolting.

The stress-strain curve consists of two parts: there is the elastic range and the plastic range. The elastic range consists of a straight line at a particular angle. The slope of this line is known as the Modulus of Elasticity and is the amount of stress divided by the amount of strain produced by that stress. For steel, regardless of the strength, the Modulus of Elasticity is 29,000,000-30,000,000 depending on where someone feels like rounding it off to and with minor differences due to chemistry. Since the modulus does not change with hardness, all bolts behave roughly identically within the elastic range. The differences become very noticeable when comparing yield points and the strain behavior within the plastic range. Hardness makes a big difference there.

One of the most beautiful things about the design and use of high strength bolts is the stress-strain curve. Bolts are manufactured to have a hardness within a specified range which affects their behavior under load, by raising the yield point to a particular level. This also has an influence on their behavior in the plastic range which is controlled and somewhat predictable. For instance, A325 bolts have a Rockwell hardness of 25-34 C for bolts up to 1", and 19-30 C for bolts over 1" diameter. The bolts are also manufactured with a very specific thread pitch to take advantage of the strain properties of the bolts in a predictable manner.

The full pretension of a bolt is specified at 70% of the ultimate tensile strength, which is a little below the yield point. That is not a bad thing as people think. Pretensioning may even cross the yield point. That does not mean that it is anywhere near breaking. It is designed that way. The yield point is the point of the stress-strain curve where the behavior goes from practically linear (it never really is in the microscopic world) to noticeably curved. The linear area is the elastic range of the material, where releasing the load will theoretically return it to its original size and shape. Past that point, it won't return to its original length, which is not necessarily a bad thing and doesn't mean that the bolt has been damaged or that it can't be reused, within limitations, assuming that the nut will go back on. It also means that the bolt will begin to stretch (strain) more with a smaller increase in stress. Even though a bolt may reach 70% of its tensile strength at 1/3 turn past snug, it may take more than one full turn, for a short bolt, to actually break it! A longer bolt may stretch more than that before breaking. As a matter of fact, the ASTM standards for the bolts talk about rotational capacity tests that are a minimum of a full turn of the nut past snug for normal bolt lengths (2/3 turn for shorter bolts) and must show no sign of damage. It is hard to impossible to get that across to people who think that 1/3 turn will break the bolt. It should be noted though, that the strain is concentrated in the threaded section, which is only somewhere around three-quarters of the nominal diameter, plus or minus a few percent, and the actual behavior of an individual bolt will be affected by the length of the threaded section included within the joint.

The 70% point of the curve, the upper region of the elastic range, is basically linear and that means that it is also very predictable. The manufacturing tolerances (at least for better manufacturers) are very tight. The hardness of the bolt, combined with the thread pitch, create a specific and reasonably predictable strain (stretch) for a given stress (tension). The importance of understanding the curve is that there can be no stress without strain, and no strain without a corresponding stress. A full turn of the nut produces a strain (stretch) of one thread-width in length, in theory. That stretch (strain) must always correspond to a specific and very predictable amount of stress (tension). If a specific amount of stress is applied to the bolt, then there will be a predictable amount of stretch. In service, and this is also impossible to get across to people, this means that unless there is a separation of the joint, there is no important increase in stress on the bolt. (Technically, there is a slight increase in bolt tension which can range from 1% to 7% before separation depending on which study you read.) That is why the commentary to the RCSC specification says that it is assumed that if a bolt is not broken on installation, it won't break in service even if it was over-tightened. There can be no separation of a joint without there being an increase in stress on the bolt. There can be no separation of the joint unless the clamping force exerted by the bolts is first overcome by a greater force (and that clamping force can be from a few tens of thousands to over a million pounds). The tighter that bolts are, the greater the clamping force exerted to keep the joint from separating, and the greater the force, like seismic, that would be required to spread the joint. Unless the joint spreads, there will theoretically be no damage to the bolts. Without an increase in strain (stretch) there is no increase in stress. It is the tension of the bolts that protects them from damage. In a joint that is not pretensioned, or not pretensioned right, a smaller force could open the joint, which would occur unevenly and place more stress on some bolts than others and break them, or cause fatigue type loading which will eventually break them and possible other things as well.

Therefore, the stress-strain curve is the secret to understanding the engineering of bolted connections.

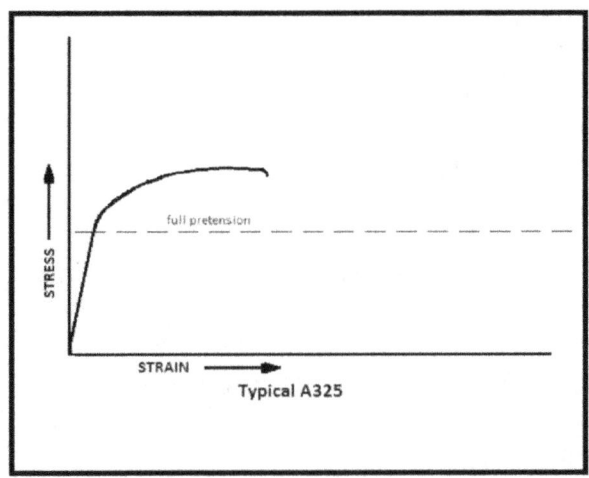

BOLTED JOINTS

The joint configuration and the loads to be transmitted through the joint, which can be impressively large with seismic acceleration, determine the bolting requirements.

There are a few categories of bolted joints, with and without pretension. Old terms for categories mentioned in old outdated manuals or specifications mention "bearing" vs. "slip-critical". Those terms still have meaning, but people using those terms often are confusing themselves, since not calling something "slip-critical" does not mean that it is not a pretensioned joint. That is why they use the more specific categories *shear, tension, combined shear and tension, snugtight (ST), pretensioned (PT), and slip-critical (SC)*.

Success of the joint begins with fabrication. Holes have to match at least well enough. The holes must have the specified spacing and edge clearances within the code specified tolerances or better. Reaming holes in the field is a reality, but that is reducing the factor of safety to a certain degree (not necessarily by much for very limited reaming) by removing metal. The AISC and RCSC requirements are based on the assumption that a small amount of reaming and pin hammering distortion will occur (to a limited amount). Hole shape can influence the way the bolt helps to transmit loads between members. Reaming often results in a slot instead of a hole.

Joints are to transmit design loads, but without overloading the bolts beyond design tension. Good fit-up is required. Plates need to be straight to come into *firm contact* with no possibility of movement if there is a shift in load. 1/20 is the limit of the angle of alignment for plates to be joined without stressing the bolts the wrong way. That angle is close enough to be closed with effort and transmit loads between plates, but if that is combined with other problems the joint might not perform as well.

There are standard holes, oversized holes, and slotted holes. Standard holes do not require washers unless specified or designed as such, or a part of a bolted *assembly* (TC) manufactured and tested with a washer. Oversized holes require washers to develop the full pressure effect (clamping force) of the bolt on the plates in a manner consistent with design assumptions.

Slotted holes require washers, and are also divided into short-slotted holes (up to 1/4") and long slotted holes (>1/4"). The performance is also affected by the orientation of the slot in relation to the direction of loading. If the long dimension is *perpendicular* to the load, there won't be any more slip possible than with a standard hole. If *parallel* to the load, it can be designed as *slip critical*, where there is a friction force preventing movement of the joint in any direction. They can still affect relaxation either way.

Slip critical joints are special pretensioned joints which only differ from other pretensioned joints in that they depend on surface preparation to develop a friction bond between plates. (It should be noted that *slip critical* is only meant to resist shear, and has no meaning with a tension load, since it is meant to resist lateral movement from sliding of the plates and obviously the coefficient of friction has little to do with separation of a joint perpendicular to the surface of the plate. That is part of why the old commonly used joint terms are deceptive.) The surface must be inspected **before the steel is erected and fit together in place.** It **cannot** be verified later without removing the members. The friction bond prevents slipping of the joint when even 1/16" of slip is bad. That may not seem like much, but it can add up to more than it seems when multiple joints are considered, and it takes very little distortion to break a window or prevent a door from closing. Slipping even a little under fatigue type load reversal can cause failure, and so slip critical joints are specified for vibratory loads such as machine bases and similar situations.

Slip critical joints used to be the only joints that can be considered to share a load with a weld, which at that time automatically excluded A307 bolts since those can't be pretensioned to a degree worth considering. Now, still only in shear and under very specific joint conditions, any type of bolt can be considered to add a little resistance to certain longitudinally loaded fillet welds, with the amount contributed being 50% of the "available strength" of the similar bolted joint. Slip-critical joints using high strength bolts obviously still provide a greater contribution to the joint from the greater shear capacity provided by the friction bond, and depending upon the coefficient of friction. The frictional resistance provided by a joint is related to the coefficient of friction of the surfaces of the plates. The coefficient of friction is a number normally less than one which relates the force of the friction in the plane of the surface as a percentage of the force exerted upon the surfaces. The pressure created by the bolts on the surface of the joint is multiplied by the coefficient of friction to give the amount of extra shear resistance created by the bolts, and that means that greater pressure produces greater friction in a particular joint.

Welds and bolts behave very differently under load. Welded joints have far less ductility than bolted joints. Bolts have the ability to stretch under axial tension, and deform somewhat under shear, whereas welds aren't going to stretch to any noticeable degree under load in a bolted joint. If the bolts have to be strained (see the first section discussing the stress/strain curve) to absorb an amount of stress, the welds will have to take the full load and would break the bolts decide to share a significant portion of the load. Obviously, pretension would greatly improve the performance of a joint in combination with welds by creating a stronger friction bond to take some load off the welds. The clamping force of a slip critical joint would take some of the *shear* load from the weld. It would make a stronger connection than other joint types because of the greater shear capacity of the joint, of which 50% is being considered. (Obviously, this only applies to a *shear* load, since friction has little meaning in tension, which must be designed as either welded **or** bolted.)

Fit-up and alignment, which includes flatness or straightness of parts, is critical to the performance of the joint. The tension produced in a bolt is only a means to an end. The point is the *clamping force* on the plates which produces *pressure between the plates*. If the tension on the bolt is wasted closing a gap (such as is common with poor fit-up and alignment and angles worse than 1/20), though there is tension on the *bolt*, it will not support the *joint*. Though the bolt at that time may be exerting pressure *on the outside surface of the plates*, if the pressure is not produced *between the plates* it is wasted.

A490 bolts have difficulty in joints designed as snug-tight, especially when there is any part of a load that is tensile, even if it is static. If a joint requires an A490 to resist the forces, then a snugtight one would be for shear only, where tension makes little difference. In tension, a snugtight A490 with an identical modulus of elasticity would not resist the force any better than a cheaper A325, and if an A490 was required it would basically not perform as such in tension if only snugtight. Snugtight is for simple shear where some slight slip is allowable. That isn't the only factor excluding a snugtight A490 in tension, though (i.e. nut thread yielding behavior).

It is just a fact of fabrication reality that if a joint has more than 3 bolts, it is likely that one or more will already be in bearing on installation. There may be much work with hammering pins and yanking bars to get all the bolts in place. Damage to the threads of the bolts on installation must be avoided. Distortion of the holes can affect performance of the bolt. Reaming can even become so severe that welded plate washers are necessary to provide adequate bearing. A small amount of this kind of work is considered normal and the AISC and RCSC requirements take that into consideration, but that assumes that it is a very small limited amount, and that it does not affect every bolt.

As one moves from the cartoon world of plans and designs into the real world, sometimes fit-up requires shims. Sometimes there are even finger shims. As stupid as they look, they work if they have been factored into the design. Plate thickness and bolt diameter affect the force concentration and area of influence of a bolt. This affects the way that the bearing pressure is distributed over a joint. Pressure is exerted in a cone shape through the cross section on the plate around the bolt and so the area of influence and the concentration of force within that area are influenced by the thickness of the plates. It is apparent that there could be an optimal bolt spacing for maximum effect, and for a certain thickness range and bolt diameter that is considered to be a little more than three bolt diameters with standard holes in common situations. It has also been observed that the ratio of plate thickness to bolt diameter has an effect on the stiffness and ductility of the joint, affecting the way the pressure from the bolt is distributed and how the plates behave, especially for slip-critical joints, and for some designs there can be an optimum ratio. These factors can mean that for some joint designs large areas of the plates may be transmitting much less of the bolt pressure than other portions and are there for other reasons. Loads may have an uneven distribution across the area of a plate. Even finger shims can still transmit the load of the bolts by extending over the area of the bolt's influence, with less than 1/4 of that area lost from the open slot. This loss affects the way the cone of influence extends through the joint and the concentration of the force. If considered by a design which compensates for that loss, that small loss around an individual bolt can be made insignificant and the joint will perform acceptably, *as long as it is pretensioned correctly, i.e. slip critical.* Any loosening or shifting of a joint with finger shims is even worse than for other joints since the joint is basically already separated and has potential for bending forces to be introduced on the bolt. Remember that pressure is the **only** thing holding a finger shim in place. Any slip in the joint could move the shim (since it contains open slots for insertion after bolts are installed).

There are also limitations to the *area* of bolted joints. The area can affect reliability through increasing the risk of corrosion, which can create a significant degree of pressure (expansion) and damage. The length of a joint affects the load bearing capacity of the bolts, and if the joint beyond over a certain length a strength reduction factor (20% per bolt) is applied to the bolts.

Looking at the strength of a connection, it becomes apparent that a certain number of bolts of a large enough size for a given connection would develop a potential load transfer (for instance shear) capacity equal to the limit of the strength of the welds holding the lug plate to the side of a column, i.e. a given number of inches of fillet weld of a given size. Assuming that it could be necessary to go maybe beyond that for reliability of the bolted joint, it can be seen that going further adds nothing to the strength of the connection for a given design above a certain amount (consider block shear rupture or differential yielding which can make the rest of the joint meaningless and can be made worse when one thinks that more is always better without considering that to add one must also remove).

Why might this be an issue worth considering, even if one is not the engineer responsible for the design of the structure? Because people have the impression that more must be better, and that adding more on top of a serious defect of fabrication or erection damage will allow one to ignore the problem. Adding bolts, or increasing the size of a joint beyond reasonable and practical design may provide little reinforcement considering real world phenomena, which can cause yielding or failure of a part of the a long joint before the load has been fully developed in the rest of the bolts further down the joint (uneven load distribution), hence the strength reduction factor per bolt per bolt (20%) when the joint exceeds 50 inches in the direction parallel to the load.

This misconception can be compared to the similar myth that potential serious defects in a weld done improperly can be mitigated by making a bigger weld or a longer weld. Just because a defective weld is larger than specified does not mean it can take a greater load before failure. Any defect that can cause failure of a weld would still exist and still be a defect, and is a crack waiting to happen. (See AWS D1.1 which does not allow welding over a crack for this reason, since the crack would spread to the new metal.) You can't fix a cracked weld by making a bigger cracked weld. That only makes it worse. A cracked weld has zero strength, and even a million times zero is still zero. The same misunderstandings can and do occur in any area of steel construction, and if one is not knowledgeable and observant one could be deceived into passing a defective joint thought to be improved with an unapproved alteration that causes it to behave in a manner inconsistent with design assumptions.

That does not mean that there is no benefit to having a greater number of smaller bolts, since each bolt would take a smaller portion of the total load with better load distribution if designed properly. But advantages of a design method always come with disadvantages. Larger bolts mean larger holes and a plate must have enough space to fit them, even though a larger bolt may be capable of taking a larger load. Multiple lines of smaller bolts may be easier to fit in a joint of some configurations, but the ratio of plate thickness to bolt diameter will influence the stiffness and ductility of the joint, and there are limits to how far one can go before one is making the problem worse by changing sizes and quantities and spacing.

CARELESSNESS CANNOT BE EXCUSED BY OVERDESIGN! The rules have been developed and proven by experts who know how important every one of those rules can be. They exist for very good reasons. Don't become another example.

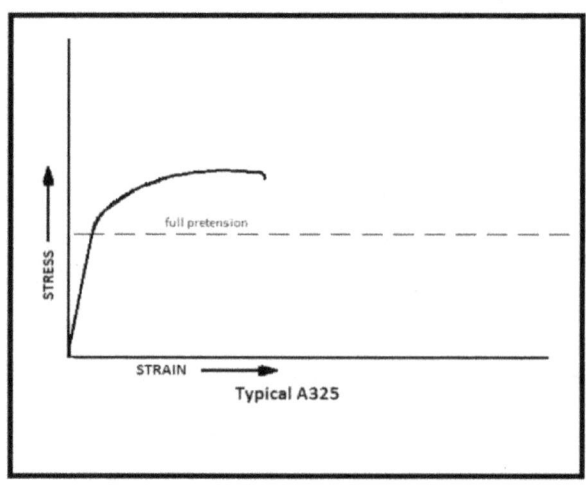
Typical A325

FORCES ON BOLTS

It's not easy being a bolt. They do the best they know how, but we must try not to make their job harder than it already is. The designers, ironworkers, and inspectors must give the bolts the best possible cooperation and treat the bolts with respect.

It is important to understand the forces acting on the bolts so as to understand why they should not take some shortcuts and why the engineer shouldn't and probably won't accept what they try to do.

One of the things expected of bolts is that they take loads without breaking, and they do that in several ways, with several types of loads and joint configurations. A few of the considerations are: shear, tension, or shear and tension, residual torsion, friction effects on tightening, impact and shock from tightening, prying action from some joint configurations (tension), bending from deviations in alignment, eccentricity and uneven loading on a joint. The bolts are designed with a certain amount of ductility to handle these situations without breaking, as long as the actual loads on the bolts are limited to only the factored loads as designed and not made worse through other problems. Residual stresses from manufacturing are present in the bolt, and slight variations in the machining or rolling of the bolt have a measurable influence on performance.

Some loads are important to understand when installing, fitting, or pretensioning, as they can increase the stress in a bolt beyond the design pretension. Torsional stresses are added to the tensile stresses on installation, and that temporary rise in stress has been known to occasionally exceed the bolt strength, although, after installation and tightening of all the bolts in a joint, residual torsion stress can be neglected because it is reduced to zero (sort of) as tensile forces reach their limit if the bolts stretch too much and the joint begins to separate. In service, after tightening, the residual torsion stress is not considered additive with the tension stress for that reason and is neglected. Therefore, if the bolt did not break on installation it is not considered a risk, and has no further effect on bolt strength apart from some possible reduction in ductility from strain hardening to a minor degree. Residual stresses can have an influence on hydrogen embrittlement.

There are ways that the torsional stress is increased on tightening. If a bolt has rust or thread damage, it would take more torque to produce the same tension in the bolt, thereby increasing the torsional stress. Thread damage can especially occur with poor fabrication, which may require forcing the bolt through the holes. A bolt may be improperly lubricated, which obviously affects friction, which is especially a concern when TC bolts are being used. *Only the manufacturer can relubricate a TC bolt. If a bolt must be relubricated on site, the manufacturer must be contacted from proper instructions. If procedures are not followed exactly, then one can't blame the manufacturer for any failures. The manufacturer can only guarantee their bolts in the condition they are in when they leave the factory. The lubricant must be the exact same type and in the exact same quantity. Even too much lubricant makes behavior less predictable. They are designed to snap off at a certain torque which is usually supposed to produce more than the minimum tension. If it snaps too early it wound not have the required tension. If it snaps too late it the tension may reach a much higher level and affect bolt ductility and cause uneven loading among bolts with some much tighter than others. Too much lubricant can also increase resistance, especially in cold weather, and so the lubricant can produce the opposite effect under some conditions. One can never guess, and so the procedure must be obtained and followed exactly, or the bolts should be sent back or replaced.* These effects are observable on a Skidmore, and one can see changes throughout the day as temperature changes.

Impact wrenches are a fact of life, as well as slug wrenches. I would prefer to see a steady pressure used to turn a bolt since the residual torsional stress from an even pressure is far less potentially damaging to the ductility of a bolt than the shock stresses from impact. The loss of ductility is a fact, but it is neglected from a design perspective and the RCSC Specification assumes the use of an impact gun. The bolts need all the ductility they can keep so they can work as a team. The extra 30% remaining tensile strength is usually enough, but if impact is combined with other horrible problems one could expect to see more broken bolts than if a gentler method was used. But, means and methods are up to the contractor unless the engineer specifically forbids torturing the bolts to death with shock stresses. Any pretensioning method they choose should have been approved by the engineer before they begin.

Shear occurs in the bolts when they are in bearing on the edges of the holes, either from slip of the joint or from fabrication and erection tolerances which are a normal occurance. If a joint has three or more bolts, one or more will likely be in bearing on installation as a fact of life.

Beam end connections are a typical example of where shear joints occur (the welded plates are often called *shear tabs*). When no seismic forces of any significant extreme are to be resisted by that variety of connection, and it is only required to hold a small amount of static weight with no bending, it can be referred to as *simple shear*. Only the shear strength of the bolts is being considered in design and these are often designed as *snug tight*. In simple shear the amount of pretension in a joint has little or no effect on the performance of the bolt, but pretensioning may be specified for other reasons such as loosening or vibration. The case of simple shear is the one case where A490 bolts may be installed as snug tight, but that is not common (see previous section). Even with simple shear connections, slip critical joints may be specified for some reasons such as frames which must maintain an exact alignment for things to work, such as around a door, or around large panes of glass that don't take bending forces well. Slip-critical joints have greater shear resistance than other joints, through the friction created between the plates which must be overcome also before the joint can slip and place bearing stresses on the bolt.

There is a problem that can occur in the extreme case when a joint slips during erection, and not all the bolts have been installed or tightened. Bolts can slip into bearing before tightening, and that may occur unevenly because of fabrication tolerances, i.e. one bolt of the group, especially with some joint designs and thickness of material. In extreme cases the one bolt can then become "trapped." A bolt can also become significantly bent from the ironworkers trying to hammer a bolt through misaligned holes. What happens with "bolt trapping" is that the bolt is bound between the plates in shear, and when one tries to pretension the bolt, the shear of the plates digs into the threads, with bolts either bent or straight depending on the severity of the situation, and the section of the bolt that is "trapped" cannot strain properly. All of the pretension is concentrated in half of the bolt on one side of the shear plane, rather than the stress being distributed throughout the whole length of the bolt, as though the plate has become the new head of the bolt. The tension goes to the face of the plate and is not transmitted through the remainder of the bolt, and therefore no clamping force exists on the joint. Turning the nut only pulls at the plate, not the head of the bolt, which causes the rest of the bolt to remain loose. Trapping has the effect of shortening the bolt and placing the effective grip outside the shear plane. The tension is then concentrated on only a small section of the bolt, the exposed section able to yield, and that section could become overloaded and fail with a given amount of rotation, even though the rotation and pretensioning may have been performed to spec. That bolt is worthless. It does nothing to hold the joint together. It is not the tension on the bolt that is the point. That is only a means to an end. The point is the clamping force that makes the plies act as one to transmit loads. That requires force to be applied by the bolts *through* the joint.

Since threads can influence the potential for trapping, threads can be excluded (marked on the drawings with an X) from the shear plane to help reduce the risk. When that is not considered a significant risk (i.e. the ironworkers know what they are doing and fabrication tolerances have not been exceeded) threads are usually not considered to be a problem when included (marked on the drawings with an N). One can sometimes see, when ironworkers attempt to remove a bolt that is trapped in shear, it may be impossible to remove the bolt without cutting it out with a torch. When they do come out, they are often bent. There is a reduction in shear strength because of the reduced section of the threads (20-25%), but that is often neglected with proper design which factors in the reduced bolt diameter at the threaded section.

A similar situation occurs at anchor bolts, which is one reason why anchor bolts are not tightened past snug unless a proper length is left unbonded and free to strain. The part of the anchor embedded in the footing is unable to stretch so that all of the strain would be concentrated in the portion exposed, rather than having the stress distributed more equally along the length of the anchor.

Bolts and joints often experience combined shear and tension, as well as bending. If such a situation is likely, then a joint will be designed for the combined forces using the "elliptical solution" or some other such formula. Tension on a joint will influence the shear resistance of a slip-critical joint. The greater the tension on the joint (not bolt tension), the lower the slip resistance in shear becomes, since the pressure between the plates which produces the friction is reduced by the amount of tension trying to separate the joint.

Take a look at Math-like Thing #3-1 below. This is only a part of the formula used in the design of joints for slip resistance with any part of the load being tension, including combined shear and tension, but it is the only part which is necessary for this discussion. One may think that having a bolt too tight is dangerous under combined forces, and by too tight I mean that the pretension meets the bare minimum required. Looking at the formula, one could recognize that tension is just as critical to the performance of the joint as in any other case. The things one should note about this formula are: 1) this portion of the formula is always less than one and must be more than zero if the joint will take any load, 2) the required tension is in the numerator, 3) one must make the actual tension in the bolts greater than required if one wants this fraction (which is subtracted from one) to be as small as possible to improve the joint reliability, 4) as this fraction approaches one the slip resistance and reliability of the joint approach zero. Greater pretension means greater reliability by increasing the denominator and reducing the value of the fraction, which is a ratio of required tension to actual tension for a group of bolts. As with all cases, it is more tension which protects the bolts and the structure. That formula assumes that the minimum pretension, at least, is met.

There is a number in this – 1.13 – which is referred to by some as a reliability rating. This number is affected by the pretensioning method. 1.13 is the lowest value one can assume, and in reality it only exists under perfect conditions. This number could be considered to be the reliability of the inspection more than of the bolts because it is so dependent on the inspector making sure things are done right and are done in such a way that there is a great degree of certainty that the minimum is met. There is in this number a probability that things were done right and bolts were tightened to the minimum tension, but that probability drops to zero if the code required minimum standards are not met. 1.13 is considered the lowest, but is the only value assumed in this formula, and it especially applies to torque-oriented methods, with turn-of-nut having a possible reliability much greater than that but only considered 1.13. There is *no* reliability with failure or negligence, and the contractor is assumed to and expected to be trying to go far enough past the minimum to be certain that the minimum is met.

That formula deals with multiple bolts in a slip-critical joint. Any joint other than one meeting slip-critical requirements would not have the friction necessary to make that formula mean much. For an individual bolt the formula simplifies to the form of Math-like Thing #2. The number of bolts in that case is one and one can't speak of an average, such as 1.13, because an individual bolt can only have one value with no spread and it either meets the minimum or it doesn't.

Joints in Tension:
as designed slip resistance

$$\left(1 - \frac{T_u}{1.13\, T_m N_b}\right)$$

T_u = total required strength in tension to resist the factored loads on the joint

1.13 = theoretical reliability factor for less reliable torque oriented methods

T_m = specified <u>minimum</u> pretension as designed, since one can't count on it averaging to anything more than the minimum

N_b = effective total number of bolts rsisting the total load on the joint

MATH-LIKE THING #3-1

For an Individual Bolt

$$\left(1 - \frac{T_u}{T_m}\right)$$

MATH-LIKE THING 3-2

When one looks at the individual bolt, it is even clearer that tighter is far safer than looser. One does not have to worry about damaging the bolt, and even combined forces are not so much a bolt problem as a *joint design* problem. Even though tension may not affect the *shear strength of the bolt*, it *can* affect the way the *joint* resists shear.

Improper tightening sequence causes uneven tension in the bolts across the joint. Tightening a bolt is always loosening another to some degree. This relaxation can be minimized by selecting the proper tightening sequence. This is a problem with both snug tight joints and pretensioned joints, and it often happens that bolts that were supposed to be snug tight, or even pretensioned, can be turned with your fingers! Improper sequencing, especially if a joint was not properly snugged first, will cause the first bolts to relax as the load is shifted to more bolts, and the loads are then concentrated on the last few bolts only. The first bolts are then too loose to share the load. One can see this when checking with a torque wrench. The first bolt tightened with the commonly and mistakenly used circle pattern (such as clockwise) of pretensioning will fail, and even if they retighten that one the next will fail, and you can follow the same circle failing the bolts in sequence until you get to the tighter ones at the end. It seems odd that a person may know how to tighten a wheel or cylinder head on their truck, but can't understand that the same principle applies to everything bolted!

Overloading individual bolts too far beyond the tension of the other bolts in the joint is bad. Bolts are meant to work together, and so should have as close to the same tension as possible. Ductility allows a degree of cooperation and distribution of forces among the group of bolts. An overly strain hardened bolt may react differently under a certain force than a more predictably ductile bolt. That is why limitations are placed on the maximum hardness of a high strength bolt, and the bolts are given a narrow range of hardness and strength, with greater limitations on A490 bolts. The joint is designed with the assumption of predictable distribution of and behavior under load. The forces on the bolts should be understood so that one may recognize potential problems and the damage (usually difficult or impossible to spot after the joint has been fit) that can be caused by mistakes.

Anything that loads some bolts but not others is bad. This can occur from either bad joint design, or more likely bad fabrication and installation.

Anything that increases loads too far beyond design loads on individual bolts, and leaves residual damage, is bad.

Anything that causes the bolt stress to be outside the joint where it doesn't do anything, such as trapping, is worthless.

A situation that will endanger one bolt endangers all the bolts in that joint, and potentially the entire structure.

The bolts are safe if, and only if, treated right. They are not as cooperative if they are not given the respect they deserve.

Other problems which can occur with a bolt are delayed cracking, hydrogen embrittlement (a manufacturing problem), or fatigue loading (a design problem). There is also a 3% to 8% relaxation of bolt tension soon after tightening, which can potentially affect the bolt's ability to perform if not accounted and compensated for in design or applied tension. Compressible coatings affect the ability to retain pretension. Delayed cracking is partly preventable by using measures to maintain the remaining ductility of the bolt, and not overloading it by other mistakes. Delayed cracking can be influenced by hydrogen embrittlement, and so it is partly uncontrollable, but the extra 30% of the bolt's strength after pretensioning (assuming that this is a perfect world) should minimize the danger. The problem is more serious with A490 bolts because of the much lower ductility, and manufacturing tolerances are just broad enough to result in some of those bolts having a yield point closer to the breaking point, and it is the distance between the yield point and the breaking point that can allow a bolt to resist the effects of such thing as hydrogen embrittlement. But, A490 bolts will still behave in a ductile manner, even if the margin is smaller, and it is not considered a problem for steels under 180ksi.

There are limitations on galvanizing of high strength bolts because of the hydrogen embrittlement risk. The coating zinc can also penetrate into the grain boundaries increasing stress at the surface. Hydrogen is introduced in the surface of the bolt through the pickling before the dipping. The acids used in the pickling can attack the surface of the steel. The dipping then heats the bolt to a certain degree, allowing hydrogen to be absorbed more readily, while trapping it in place with the coating. The risk is somewhat lessened by mechanical galvanizing. This is a process which does not require hot dipping. It involves placing a bolt in a container with zinc dust and bombarding the bolt and dust with beads until it sticks. As stupid as that sounds, it works to some extent, and so mechanical galvanizing of bolts is allowed for some applications where hot dip galvanizing would not be allowed.

Even though residual torsional stress is neglected in design, and it can drop to zero at the point of joint separation, therefore having little *theoretical* effect on the clamping ability of the bolt, when it is combined with hydrogen embrittlement delayed cracking can result with failure of the bolt. That is something that is not easy to see before it becomes a problem.

There are situations when bending forces are introduced in the bolt. That is always undesirable, especially if bending stresses occur at a threaded portion of the bolt, which can be considered one long spiral notch. The bending forces are more likely to be introduced from fabrication or erection errors, but it is ignored if it is kept within the limits of the RCSC and AISC codes. A limit of 1/20 pitch between faying surfaces is within the range of what a bolt can take without reduction of load capability, but it is still something that should be avoided. Movement of a joint can introduce bending stresses. If a joint has a degree of misalignment, movement can be even worse on the joint. Prying action introduces bending stresses on some bolts in a joint while also increasing tensile forces.

The greatest danger to a bolt is still not being tight enough, or bolts in a group not being tightened evenly and not sharing the load. That is why undertightness is always a cause for rejection, and overtightness is not a cause for rejection, per RCSC Spec.

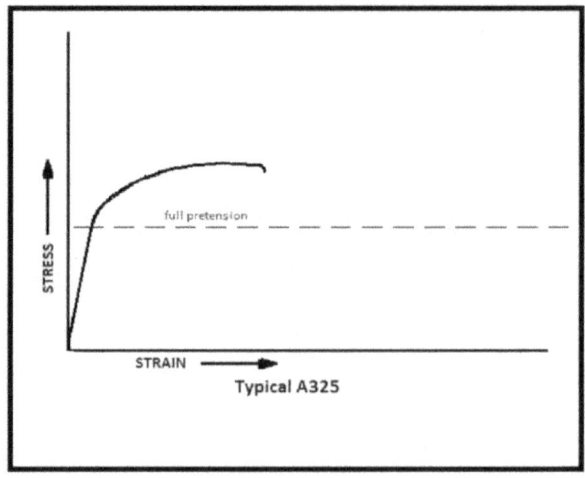

Typical A325

PRETENSIONING: where the fun begins.

If I intend to discuss pretensioning, why would I begin by discussing snugtight joints? Because snugtight is the first step in pretensioning and it is necessary to understand both the importance of getting it right and to also understand the differences.

People frequently ignore the definition of "snugtight", which does not mean "fingertight", since no such thing exists from a practical standpoint. (Yes, I have seen a sloppy engineer write "finger tight", but that doesn't mean the ironworkers should leave their wrenches at home and turn the nuts with their fingers.) Snugtight, which is the first step in pretensioning, and also the last step in snugtight joints (simple shear without tension or fatigue loading and not specified as pretensioned), must be understood, or everything after will be unpredictable.

Snugtight is the "full effort of an ironworker with an ordinary spud wrench." That force can be significant if full effort is actually used by large people. For a bolt less than 3/4" diameter, it is essentially the same as full pretension (speaking of an A325 of course), and for a 1/2" bolt full effort (actual full effort and not what they try to get away with calling full effort), which puts a person's full weight on the wrench (easily several hundred foot-pounds), it would break. For that reason, 1/2" bolts are usually specified as snugtight only, and even then it wouldn't take a full effort to produce full pretension (12 kips, which is only a short step away from the breaking point of 16 kips, which is a narrow margin in terms of a spud wrench).

Snugtightness doesn't have an important effect on shear strength, since the load in that case is perpendicular to the line of tension, which in an A325 bolt is enough in cases where there is no risk of separation of the joint (such as anything that would produce tension at the joint). A490 bolts have an extra restriction placed on them, as will be discussed shortly. Sometimes one may be lucky enough to have a contractor using TC bolts in joints designed as snugtight (a waste of money), and then snap them off, which, even if the bolts snapped off at a lower tension than they normally should and wouldn't have passed for a pretensioned joint, they are certainly beyond the minimum for snugtight. If an engineer writes that the bolts only need to be snugtight, and the contractor chooses to use F1852 bolts, should the engineer get upset if they didn't snap off one of the joints in the structure, even though it is snugtight? TC bolts are much more expensive.

When any part of the force to be resisted is tensile, A490 bolts must be pretensioned. They are only allowed to be snugtight when there is only plain shear on the joint. When one looks at the modulus of elasticity of different bolts, one notices that they are so close that they can be considered identical, depending on what place one chooses to round off the numbers. This means that a given amount of stress will have an equal amount of associated strain for both A325 and A490 bolts. A given rotation will produce the same pressure within the elastic range. In snugtight joints, an A490 won't keep the plates together any better under tension than an A325 will. It would be a waste of the strength of the A490. They have a much higher shear strength than A325 bolts. When fully pretensioned they have a much greater tension than an A325 because of the higher elastic range. When snugtight, their only advantage over A325 bolts is in shear. If one considered snugtight to be, say 10% of the full pretension for some larger bolts, or any percentage, then a snugtight A490 would be at a much lower percent than a snugtight A325, and hardness can affect the distortion and yielding of the threads at the nut. But, those points aren't the only issues and there is more involved than is being discussed here. Friction resists the torque from tightening at first, but after tightening that friction resists loosening. In shear, bolt tension has little effect. When a joint experiences tension, the tightness, and the continuing tightness over time without loosening and relaxation, makes a difference. Tension in the bolt and pressure and distortion of the threads lock the nut in place, as well as friction.

The purpose of pretensioning is to prevent separation of the joint, so loads can be transferred through the joint as if the plates were one piece. Much already discussed regarding the stress/strain curve bears repeating. Pretensioning induces a tension in the bolt of a certain minimum amount. Turning the nut a certain amount strains the bolt a certain amount, which is very predictable through the elastic range, and that creates a certain amount of tension in the bolt. The joint can't separate unless that tension is overcome, such as in the case of seismic acceleration. The tighter the bolts are, the less likely they are to break. No matter how strong the force becomes, if the joint does not separate, then there is not a real increase of stress on the bolts in the joint. If the joint does not separate then there is no increase in the strain (stretch) in the bolts. If there is no increase in strain, then the stress/strain curve shows you that there is no increase in stress (tension) in the bolt, except a few percent right at the point of separation, varying in value between 1% and 5% depending on the study you look at.

Many people seem to think that the forces, pretension and loading in service, are additive, as though one would look at the 51 kips of tension in a bolt (70% of the theoretical ultimate tensile strength of a 1" A325 bolt) and then add an earthquake load of so many more kips and think that it would break the bolt when it already had so much force in it. There are many people who think that the bolts would be less likely to break if they are looser! The reality is that the forces are not added. It is the difference between the two forces that matters, with the larger number needing to be the bolt pretension, and the clamping force being the load to be resisted subtracted from the bolt pretension. As shown in the previous chapter, the risk to the bolt is in the ratio of calculated load to pretension. There can be no damage to the bolt unless the joint separates, as noted above and is well proven. The joint cannot separate until the pressure between the plates is first brought to zero. One force holds the joint together and the other force tries to pull the joint apart. Only after the clamping force has been overcome can there be damage. At the moment of separation of a joint the tension in a bolt only increases 1% to 5% depending on who you ask. The tighter the bolts are, the greater the force would have to be to overcome the bolt tension. The chance of a joint successfully resisting a load can be viewed as subtracting the tension of the load from the tension in the bolts. This should always be greater than 0, and much greater to provide factors of safety. It is the same with shear loads on a slip-critical joint. The tension in the bolts creates pressure and friction between the plates. The frictional resistance must be overcome by a greater force to cause the joint to slip. The friction in the joint is directly proportional to the bolt tension which creates the pressure, and the greater the tension, the greater the slip resistance becomes, and the greater the force must be to cause the joint to slip. For any type of joint, the looser the bolts are the smaller the load has to be to overcome them. Therefore, it is the tightness of the bolts in pretension that protects them, even if many people, ironworkers and inspectors both, think that they know more than the engineers and researchers who wrote the code requirements for high strength bolting. The bolts are designed to be tightened to the tension specified in the RCSC Specification, and the stress/strain curve is the key to understanding these requirements. The problem is that most people in the business don't correctly understand the stress/strain curve, if at all. There are reasons for these rules, and it is stupid to endanger people's lives and property by disregarding them.

It should also be repeated that the methods of pretensioning are not the point. Turning a nut is only a means to an end. It produces strain in the bolt which equals a certain amount of tension, but that is also only a means to an end. The tension in the bolt puts pressure on the plates (and by plates I mean anything that is connected by the bolts), but that is also only a means to an end. The real goal is the pressure created *between* the plates. Unless things are done properly with all bolts in a joint being properly snugged first before any bolts are pretensioned, and proper fit-up methods are observed, then only the last bolt tensioned could be considered tensioned since the others lost their effective tension in the process. It is not the tension in one single bolt which is the point, but the *total* average effect of all the bolts working together.

Any tension applied to a bolt to draw the plates together in a joint with bad fit-up is to be considered subtracted from the apparent pretension, which is only an illusion, which can vanish if there is a shift in the frame of the structure. There may be a certain amount of tension created in the bolt, but if the joint had not been snugged or aligned right first, it does not create the required pressure *between* the plates.

There are different ways to go about creating the specified tension in the bolt, but it is all "turn-of-nut" no matter how you look at it. "Turn-of-nut" includes tuning the bolt while holding the nut still. Turning the nut is preferred, but it is left up to the contractor to choose the best method for an individual joint. There are times when the clearances around the joint make turning the bolt the best option. The effect is the same, but turning the bolt may take more torque because of friction on bolt threads in bearing with the sides of the hole. Turning the nut, if done right, does not turn the bolt in the hole and therefore takes less effort for the same effect. If the bolt is turning with the nut, that is another problem.

The biggest and most destructive mistake when determining how to produce this clamping force is to think that it centers around torque. We're not talking about wedge anchors here, which use a certain torque to jamb the sleeve into the sides of the hole. Torque and tension are two different things which have no predictable relationship, although people try to approximate it through formulas that don't account for reality. In the case of this discussion, no assumptions can be made without the required testing, no matter how impressive mathematical formulas may appear. Torque in itself is irrelevant. Torque is never the point. Torque is only a tool and never an end in itself. Torque (turning force) is obviously *involved* whenever something is turned, but the joint is designed around tension, not torque.

All pretensioning methods are turn-of-nut. The different flavors discussed in the RCSC Specification are actually methods of *verification* and assurance that the nut has been turned, i.e. ways to know that it's been tightened to the specified tension.

Because people, including inspectors, believe that torque is the end, instead of a means to an end, they forget that it has no meaning if it does not turn the nut enough to produce the required tension. It does not matter how much torque must be applied, theoretically. (Of course torque stresses can damage a bolt if, for instance, a rusty bolt with a high coefficient of friction is torqued to produce the required amount of turn, and residual stresses can result in delayed cracking. But that is neglected by design. If a bolt is that bad, it should have been discarded and it would be a violation to use it anyway.)

Look back to the stress/strain curve. (I can't emphasize enough how important that is since it is the key to understanding bolting and the engineering of steel structures in general.) There is no stress without a corresponding strain, and no strain without a corresponding stress. The stress (tension) is what we are interested in. Even tension is a means to an end, that end being the clamping force and pressure between the plates. Strain is a means to an end, and also proof that the stress has been produced. All turning, whether it involves TC bolts, torque wrenches, or measured turn-of-nut, produces a certain amount of strain (stretch) for a certain amount of rotation, which is predictable after testing, *but only after testing*. Assumptions have no place here or in the codes.

The common number for usual bolt lengths encountered (2-4 times the diameter in the grip) is 1/3 turn past snug. The thread pitch causes an exact amount of stretch for a given bolt -- always.

The variable is the amount of force required to get that 1/3 turn. There are geometric variations which exist from manufacturing tolerances which can make a noticeable difference, even from the same manufacturer.

In theory, the chart people look at when trying to proceed without the Mandatory Pre-Installation Verification testing does not take into account the thread pitch/diameter ratio. It only considers *diameters of length*. If one looks at *diameters of length* and ignores actual length, then the thread pitch makes a huge difference in the required rotation to produce the required tension. The different diameters have different thread pitches, with the number of threads per inch decreasing as diameter increases. This would give the appearance of a greater strain for a given rotation, for instance one full turn past snug, and that would be true if one were measuring the lengths in inches rather than diameters. But, *diameters of length* are what is being considered by that chart. The important consideration for this discussion is the ratio of thread pitch to diameter and the percentage of the diameter of length taken by one thread. For example, a ½ inch bolt has thirteen threads per inch, and a 1 inch bolt has eight threads per inch. The thread pitch on the 1 inch bolt is much larger, but it takes up a smaller percentage of the diameter length than the thread on the ½ inch bolt. A ½ inch bolt has 6.5 threads per diameter of length, while the 1 inch bolt has 8 threads per diameter of length – a significant increase when considering lengths measured in diameters. This can give the impression of the smaller bolts being stiffer and the larger bolts being less stiff. When considering diameters of length, therefore, the rotation required is different for each diameter and in this specific case a given rotation would produce a different percentage of the required tension (70% of tensile strength). This would make it appear under this specific case that a larger diameter bolt requires a greater amount of rotation than a smaller diameter bolt. This does not apply when one is considering a fixed length, such a one would find in a specific joint.

Other considerations which cause deviations from theoretical elongation are the yielding of the nut threads, which has the effect of increasing the apparent bolt length and reduces the tension produced by a given rotation, and the huge variation in the degree of snugness of individual bolts in a joint, which means that each bolt begins rotation from a different point.

One can be certain that with a given amount of turning (advancing the nut along the threads) there is a given amount of tension. [One should remember that tensioning into the plastic range is considered acceptable by code, since 1) too tight is nowhere near as bad as too loose, 2) the yield point is known from the mill certs and the yield point is theoretically just past the pretension point and crossing that proves that the specified tension was reached, and it is not cause for rejection as would be the case for bolts not being tight enough.] The nut can't advance along the threads unless the threads are exposed (or pulled out of) the joint (assuming of course that the joint has been properly snugged and the plies are in firm contact first, i.e. the nut can't go any further than the face of the plates unless there is a very severe problem with the holes, such as deformation, which should not have been allowed to proceed to that stage.) The thread is "pulled" out of the joint and exposed by the tension (stress) induced by the turning of the nut (which equals a certain amount of strain.) As an example, consider a 1" diameter bolt which has 8 threads per inch. One full turn of the nut past snug will stretch that bolt theoretically 1/8", assuming that there is no possibility of movement of the plates, and neglecting any yielding in the nut and bolt head. This is in theory only, for the purpose of illustration. It won't actually be the full thread pitch and the true elongation depends on other factors to a small degree. One could look at the stress/strain curve for that bolt, with consideration of the length of the bolt within the grip, and see that 1/8" of stretching produces an exact amount of tension, or equally, that the tension involved will produce exactly 1/8" of stretch for that specific bolt.

This redundancy in this text is important. It is necessary to understand the concept of turn-of-nut completely in every way, or one will not understand the principles of the other methods correctly, since all methods are actually turn-of-nut. It is necessary because of the failures of comprehension that are common enough to be called normal, even when people have taken classes in this topic.

One can also see from the stress/strain curve it can take up to several full turns past snug to break a bolt, depending on length, and bolts are tested per ASTM for rotational capacity with no sign of damage or adverse effects. The bolts are tested at far greater rotation than is usually required in the field. When a contractor whines about not wanting to turn the bolt even a minimum of 1/3 turn because it will break if they follow the code requirements, it is because they have never bothered to learn ANYTHING about bolts or about code requirements, and they just proved it. Why complain about 1/3 turn? Bolts at one time were turned far more than that. Inspectors also can fall into that category as well. That is why I am discussing this at this length and with repetition, since they had obviously missed it somewhere even though it takes very little effort to see all of these points in the commentaries printed in the AISC manual or RCSC Specification. One can see from this that 1/3 turn is nowhere near the breaking point, and it could take up to several FULL turns to break a bolt, corresponding to a stress far above the minimum required "full pretension." For short bolts it will take less turning to reach the breaking point, and for longer bolts it will take more, but one should certainly not be too worried about one full turn past snug. The particular amount of turning to produce the necessary tension is influenced by the length of the bolt within the grip of the joint, meaning the part of the bolt between the plates (assuming proper installation without trapping), and the ratio of threaded to nonthreaded shank in the grip, and the 1/3 turn, which is quite common, is only considered good up to somewhere around 4 bolt diameters of length, depending also upon other factors already discussed. A range of 4 bolt diameters is actually a large spread for a single value of rotation.

It should be kept in mind, that just because 1/3 turn could produce a force of 70% of the tensile strength of the bolt, that does not mean that 2/3 turn would produce 140%, as some people believe if they don't look at the stress/strain curve. This is because once the yield point has been crossed the stress/strain relationship becomes <u>nonlinear</u>. A given amount of stretch will produce less stress in the bolt, so a bolt can be rotated up to several full turns (but that isn't a good idea) before breaking, even though 70% of the tensile strength was reached at 1/3 turn past snug. This can be seen from the stress/strain curve. Also, as the bolt stretches (elongation) a point is reached where there is a certain amount of reduction in area of the bolt diameter, meaning that the diameter is shrinking. This reduces the *net tensile area* of the bolt. The diameter may be shrinking, but the amount of work required to turn the nut keeps increasing because of *strain hardening*. That is a part of the nonlinear portion of the curve and affects the shape of the curve at that point.

It must be remembered that all bolts have approximately the same modulus of elasticity and the stress/strain curve is roughly identical within the elastic range. But, that does not mean that one can assume that the amount of rotation required for one bolt will be identical to that required for another bolt. This is because stress/strain is a RATIO, and is independent of some of the dimensional factors, especially bolt length between the plies of the joint. To be more precise, the stress/strain curve refers only to the properties of the *steel* that the bolts are made of, and not necessarily of the bolts themselves or of a particular application of those bolts. Therefore, it can't be assumed that 1/3 turn past snug will result in the desired tension unless testing was performed. A longer bolt has more metal to stretch. Once the bolt has been tested to determine requirements (and that must be done for every set of bolts from every different lot, heat, length, diameter and so forth), one can think of the relationship as a given amount of stretch being divided by length. For example, two bolts have a thread pitch of 8 threads per inch (1" A325). One bolt will have 4 inches of length between the plies (grip) and the other bolt will have 8 inches between the plies. Both are the same diameter, as stated. A rotation of 1/3 turn past snug for the 4" bolt is performed, and takes a particular amount of torque and produces a particular amount of tension. The 8 inch bolt is then turned 1/3 turn, and very little effort was used, and the bolt is still loose. That is because if you are viewing it as strain per inch of length, which is the reality when a lab is tensile testing a bolt, there was more metal to stretch, assuming that we are talking about the effective or apparent length since there is an uneven distribution of stress and strain along the length of a bolt, and therefore the strain per inch much less and a correspondingly smaller tension was produced for a longer grip. The 1/3 turn may have been the same, but the relationship has changed, even though the both bolts may have identical chemistry and hardness, and apparently identical stress/strain curves. Always keep in mind that turning the nut is not the objective in itself, but only a means to an end, which is to produce a specified amount of pressure between the plates.

    Here is a similar example, for illustration only and neglecting various factors and is not to be considered to be exact enough for a real bolt. Let's assume one full turn of the same bolts, which would stretch both bolts nominally 1/8 inch = .125", neglecting all the other factors for illustration. For the bolt with 4 inches between the plates (only the portion of the bolts within the grip need to be considered since anything past that is not under any tension), that would be .125 divided by 4 = .03125" of stretch per inch purely in theory, (neglecting nut and head yielding and other factors for the purpose of this discussion), which can be assumed to relate to a given stress from the stress/strain curve,. (Also, we are talking about free length, which is reduced by trapping or other damage.) For the bolt with 8 inches in the grip, 1/8 inch of stretch would be .125 divided by 8 = .015625" of stretch per inch, in theory only and neglecting all other factors of reality for the purpose of illustration only, which is a much smaller amount of strain per unit of length, and which according to the stress/strain curve (ratio) relates to a much lower stress in the bolt. It can be seen that it would take much more turning to produce the same pretension in the longer bolt, even if they are *metallurgically* identical. The real elongation would be different also because the strain in not distributed identically along each inch of length and would be less than in theory, with stress concentrated more in certain areas rather than spread throughout the whole length, and more especially if the yield point is crossed and depending on how many threads are in the grip. That is why gage length is important in tensile testing.

For A325 bolts up to 1 inch in diameter, the required elongation at the breaking point is 14%, and 14.35% above 1 inch, when measuring a particular specified gage length. The diameter which matters most is not the nominal diameter but the diameter at the thread root, which is much smaller. This is purely theoretical considering only the steel that the bolt is made with and not the actual bolt itself. That does not mean that it would take 4 turns, or even 3, past snug to reach the breaking point, since snugging the bolt properly will produce some stretch. That also does not take into account the uneven distribution of stress along the length of the bolt, which is concentrated more in the threads and is affected by other considerations. It is also neglecting other factors that cause actual elongation to deviate from theory. The amount of stretch when the effective diameters are considered depends also on the ratio of threaded to unthreaded lengths within the grip. Though nominally of the same diameter, a fully threaded bolt will behave differently than a partially threaded bolt. 2 thread lengths could be a more realistic breaking point, maybe, depending on length, and there are other factors involved in how far a real bolt will actually stretch in a real joint. One should not count on that occurring in the real world of an actual connection, and the bolt is being severely damaged before reaching that point, for more normal lengths. Naturally, also, it would make a difference if the reduction in area is concentrated in a threaded portion of the bolt or in a solid shank, with a much lower strain capacity existing in the notches of the threads. With the 8 inch grip, 14% elongation would in theory be a significant amount, although that 14% involves gage length considerations in tensile testing and the length of stretch will not be as simple as multiplying 8 by 14%, since it is not 8 inches that will be stretching 14%. (This is purely theoretical. There are two points to this example. First, that bolts of different lengths have different requirements even if all else is the same. Second, 1/3 to even 1 full turn is nowhere near the breaking point. It should also be mentioned that different diameter bolts will behave a little differently if they have the same grip length since they will have different thread pitch angles.)

When moving from the pure theory above to the world of real bolts, this force spread throughout the length may not always be spread *evenly* throughout the length. The threaded portion has a different *effective diameter* than the uncut shank. There will be a difference in the hardness of the bolt at different locations, such as at the end along with the last couple threads or near the head of the bolt, since different areas of the bolt would have different rates of cooling. Isolated hard or soft spots may exist. There will be different strain behavior close to the head of the bolt and adjacent to the unthreaded shank, which also means that the amount of threaded section included in the grip can affect the strain behavior of the threaded section. Machining involved in either cutting or rolling threads will result in changes to the residual stresses. It may also take some time for forces to equalize in the bolt after pretensioning, especially if there are such unpredictable factors as threads being locked against the edges of the holes, or being trapped in more extreme cases, and bent bolts. Also, it can be assumed that there will be a 3% to 8% loss of pretension in a short time, and that can result in a reduction to some degree of the force transferring ability of the joint, although usually neglected in design. That means that one should not be trying to get only the absolute minimum pretension, since in reality one would not actually get that.

There is a limit of bolt length at 12 times the diameter. That does not mean that longer is dangerous, but it has not been studied sufficiently to be considered as predictably reliable as the lengths considered normal by the RCSC Specification. Some lengths are only available in less controlled A449. Predictability is essential. There are also limitations to the *area* of a joint, both from a practical viewpoint and from a consideration of reliability. There are also suggestions for the ratio of material thickness to bolt diameter. (See Kulak et al, 2001). With many small bolts there are some potential advantages, such as better load distribution under some circumstances, but there are also disadvantages. One disadvantage is that anything increasing the complexity or labor involved in a joint encourages cheating and sloppy work, which decreases reliability in the real world.

Why would I spend so long on a discussion regarding length, when one could look at the table in the RCSC Specification that everyone has seen which shows values of up to 2/3 turn *for a joint with good fit-up and flat plates and perfect bolts*? Because **NOTHING** is prequalified in bolting and **EVERYTHING** must be determined by testing. Skidmore testing may show 1 full turn required for a bolt to produce the specified tension, while the contractor whines that some worthless chart says that only 1/2 turn is required, when they do not realize that the chart was shown in the Specification only as an example of the effects joint configurations and situations could have on rotation requirements and is not a prescriptive chart to be accepted without the testing that the same Specification says is mandatory. The RCSC Specification requires the contractor to use the procedure qualified in the actual testing. The charts are only useful in theoretical considerations and not to be considered from a practical viewpoint. No matter how many times you show them on the Skidmore, they may not get it, including inspectors! People don't want to understand it when understanding involves what they believe is more work. The need for testing is easily understood when one considers that different diameters of bolts have different thread pitches which affect the strain of a given rotation for a given grip length, every bolt from every manufacturer will have different variations in geometry to some degree, especially if one uses rolled threads and another uses cut threads, and there are different stress/strain curves for A325 and A490 after the yield point for an A325 has been crossed, which are not taken into account on that chart, even though the chart makes no distinction. A bolt with a greater effective stiffness may take a little less turning (for example a 3/4 inch bolt), and a bolt such as a 1 inch bolt may take a little more. The contractor will also snug the bolts to different degrees in a real joint, and that is not predictable by looking at a chart. **Don't forget the requirement for Pre-installation Verification**, which you would have read in that section before that deceptive chart that throws so many people off, as well as before the section on installation. It is never the bolt being qualified in these tests, but the technique of the contractor, including snugging. Charts relating torque to tension are not allowed by the Specification as grossly unreliable, and people still get deceived by such charts, which they are not supposed to be giving any consideration to. What appears to some to give an easy and simple answer is only an illusion and only a method to produce failure. Without determining the stress/strain behavior of an individual bolt through proper testing, any rotation chart is just a blind guess. One can see with very little involvement in testing that the actual necessary rotation for an individual bolt may be as much as twice the rotation listed in those evil charts. Testing is **always** required and **nothing** is prequalified, no matter how scientific sounding a chart may appear. People can also forget that it is not the total length of the bolt that counts but only the length of the bolt within the grip, and so the bolts must be tested at the correct grip length or testing may not match reality.

For short grips, the effects of slight variations in grip can be huge. It can easily be seen from testing that each length of bolt/grip will have a different required rotation, but the chart only lists three length groupings, and that is for diameters of length. To use a chart, there would have to be a different chart for every diameter and length, taking into account the difference between fully threaded bolts and partially threaded ones, and type, and also one for every manufacturer, heat, and lot. They try to keep a close tolerance on the hardness and chemistry and dimensions, but even that is a range. <u>Test everything and make no assumptions!</u>

    Don't forget the purpose of all of this. Stretching the bolt isn't the point. Stretch is, of course, related to bolt tension as shown by the stress/strain curve, <u>but the point of all of this is to produce the pressure between the plates of a specific value</u>. All else are only means to an end. It doesn't matter how much you stretch the bolt if that pressure is not produced between the plates.

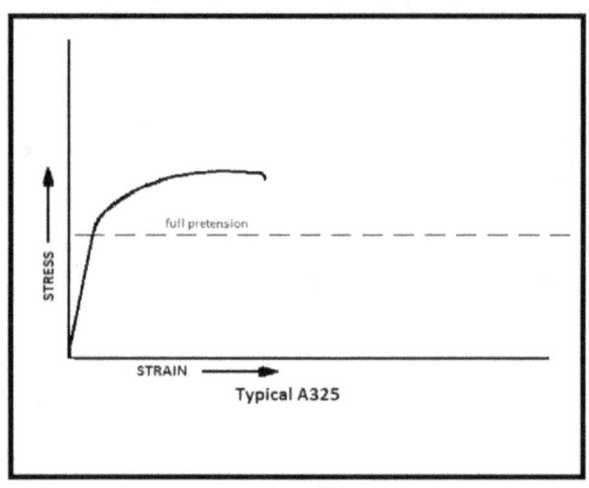

COMMENTS ON SNUGTIGHT JOINTS

No matter what type of joint is specified, it must first be snugged properly. If that is not done right, nothing after that has a good chance of being right. Snugtight is too often misconceived as meaning that there are no rules, pre-installation verification, inspection or need to even try. Snugtight joints are falsely treated as unimportant, and it is believed that they don't even need to be snugtight.

Although it must be admitted that snugtight joints are not nearly as interesting as a pretensioned joint, snugtight is the foundation of all joints and should be given great respect. It is a very uncertain state, considering it involves the least accurate methods of tightening: trying to "feel" if it is right, thereby condemning it to a spread of values ranging from minus everything to somewhat above the average assumed. That can be mitigated to some extent by use of care and proper procedures.

All joints begin at the joint, which is closely tied to fabrication and erection. Is the joint aligned within the specified tolerances? Do the holes line up without excessive reaming or distortion from pins? Are the plates bent? How about the tightening sequence? Are the threads supposed to be excluded from the shear plane(s) of the joint? Are washers specified or required for the situation? Are the bolts the right length for the joint with enough available thread so that the nut won't hit the thread runout? Were the plates clean and free of loose scale or dirt or unapproved substances, compressible or otherwise? Will the joint then be *pretensioned* after snugging? Are the other joints in the structure fit before you are going to snug this one? (That is an important consideration since tightening one joint could make other joints more difficult, or cause trapping and bending to occur as joints are shifted.)

Often, contractors don't think much about these things when a joint is specified as *snugtight*, and inspectors often don't think much about it either, misinterpreting the phrase "no <u>further</u> verification required." What that actually means is that <u>all</u> of those items noted above, plus some like sampling and testing and identification, were already verified. Then, *after* all that has been performed properly and the joint is specified as and allowed by code to be snugtight, then measurement of the *pretension* is not required. The verifications up to that point are still required, even if the joints do not require *full pretension*, and that also includes witnessing that the bolts were in fact tightened in an effective and reasonably certain manner. That method must be capable of preventing bolts from loosening, either immediately or in the future from movement or settling of the building, and also not create eccentricity in the building or improper distribution of loads from bending or undesired excessive slip, or any other consideration which may make the joint ineffective. (No matter what the joint is designed to do, it has to actually be capable of doing it.) So, the RCSC Specification says "NO *FURTHER* VERIFICATION," and does not say "*NO* verification of any kind" or imply that in any way.

What good do the bolts do if they don't hold the plates together? The bolts are required to be in "firm contact", even for snugtight joints. That can easily be done wrong if that is misinterpreted. Firm contact means that there is no possibility of movement of the plates in an undesigned way. The bolts can't even transfer shear correctly if the plates can separate and play and rotate because the bolts are so loose, or not all the bolts are tight.

Assuming the usual erection situation, the plies must be *brought* into contact for a variety of reasons, and can't be expected to just fit themselves. Accurate and careful fabrication and accurate detailing are important factors in this. This may even mean repeated tightening sequences, especially since tightening one bolt <u>always</u> means loosening another bolt to some controllable degree.

There may be straightening to be done to the plates, as long as it does not create too much undesigned eccentricity in the joint or cause other things to not function as they should. Some straightening is allowed within limits. It is not good enough just to get the bolts in the holes with the nut fully threaded. Such a happy looking condition can be deceptive. The joint must still be capable of staying together in service and transferring the required loads. The bolts must be installed without damage, such as thread damage which would prevent tightening or cause bolt trapping and shearing, or excessive bend stresses. It has been determined to be *safe for the bolt* if the plies are aligned within a 1/20 pitch. Such does not create excessive bending of the bolt. A little more than that would need a beveled washer to protect the bolt or the joint may need special correction. But, that should NEVER be thought of as implying that the plates are allowed to not be in firm contact or that a possibility of undesigned movement is allowed. Even a slotted slip joint will not do its job if the plates move out of contact. After all has been done to meet these required conditions, even though there might not be *continuous* contact at all points, there must still be no possibility of movement.

Misaligned plates or plates not in firm contact are deceptive. That is because it takes energy to bring the joint into compliance. When the bolts are used to bring the plates into compliance, then they can't be considered snugtight at that point, since the energy expended in the movement or bending of plates must be subtracted, and what is left may be so little that one can turn some of the bolts with one's fingers. The only energy that counts is the energy used *after* firm contact has been achieved and there is no possibility of movement. That is the only tension remaining in the bolts. One has not done their job simply because the plates have been brought together. We are looking for reliability even in snugtight joints.

All tension -even one pound- will cause a certain amount of strain, (although something that small would not be easy to measure,) as shown by the stress/strain curve. A principle of the stress/strain curve is the Modulus of Elasticity. Connected with that principle and the stress/strain curve is the fact that a very small movement of a joint, let's say 1/16", causes a loss of thousands or even tens of thousands of pounds of tension in a bolt. On a 1" A325 with a thread pitch of 8 threads per inch, or a pitch of .125" per thread, a movement of 1/16 inch is greater than the difference between snug and full pretension. It can also be the difference between snugtight and zero.

There are cases of slip joints with slotted holes which may sometimes be specified for some things. But, even if limited slip is designed, that does not mean that the bolts are allowed to be loose, since a loose joint can come apart, and might not transfer any load at all. There are restrictions involving the direction of the slot with regard to the line of force (unless it is designed as and meets all requirements for slip-critical joints.) The slot should also not be both a slot, either short or long, and also oversized at the same time. If the slot is transverse to the direction of load (which then would only slip under load an amount equivalent to a standard hole) then the member is probably a secondary member if it is specified as snugtight and the slot is for erection purposes, and welding may be recommended. It is unlikely that the direction of the slot would be parallel to the direction of loading for a snugtight connection unless it is for a strictly nonloadbearing architectural piece. Even then, it would only be a short slot or it could slip to the point of either nonfunction or would look stupid. Even if a joint is specified with short slots, that does not mean that there are no limits to the slotting, and one could go too far. A slot should not be fabricated or lengthened in the field any longer than designed. And, just because one joint may be slotted under some conditions and for a specific purpose, that never means that slots are allowed on other joints designed with standard holes. Oversized holes are also frowned upon for snugtight joints, since the piece could shift anywhere (except that there could possibly be cases where the engineer doesn't really care.) There are restrictions on both slots and oversized holes, which should not be considered an easy fix to a problem without the engineer performing the necessary calculations, and one should never just assume that it would be acceptable without checking.

Snugtight joints have very limited tension resisting capacity. They are generally for simple shear bearing type connections, and have restrictions on use. So, one must never assume that a joint is good enough snugged unless it is designed as such for that specific case. If a joint must be pretensioned in those situations, then the ironworkers and inspectors can't use the usual excuse that it the joint is stronger than it would ever need to be. Some such situations include fatigue type loading, some tension loading applications, seismic load resisting joints, load reversals or vibration that could rattle a snugged bolt loose, or some situations where the bolts are supposed to share a load with welds. There was a time when only bolts in slip-critical joints could be considered to share a load in shear without tension, but the requirements for load sharing have changed and in some ways are more restrictive, even if A307 bolts can now be used in these situations.

The rules that have been developed should never be disregarded. Someone should not question the engineers and scientists who made those rules, since those codes are based on experience. That experience has come from disasters and those codes have been written in blood. Many people don't understand the reasons, and disregard the rules because they think that they know better. People often use the excuse that they have never seen a joint fail after they have done it wrong, but they should not be endangering people's lives to try to save time or money. Structures are much more expensive to fix or replace and the wrongful death lawsuits are not worth it. It is cheaper to do it right the first time. Another consideration is that even if a building doesn't fall down from a joint failure or slip the building may still become unusable when things shift. It is wrong when they say that the structure was overdesigned, which may or may not be true to *some* degree, but that does not mean that it would meet code requirements. Experience has shown the necessity of the use of factors of safety and resistance factors in the calculations. These are to account for the unforeseen. These help to mitigate the flaws created in fabrication and erection, especially when things refuse to go together without bending and reaming. It is because of these design considerations that buildings and bridges stay out of the news, and why those ironworkers haven't seen joints fail (also the fact that the project already ended and they may be in another state. It is their managers who end up in court.) If it wasn't for those factors of safety, there would have been more failures. That is definite, since those disasters have already occurred and we are trying to keep them from happening again.

It is all about what *could* happen and not about the situation *right now*.

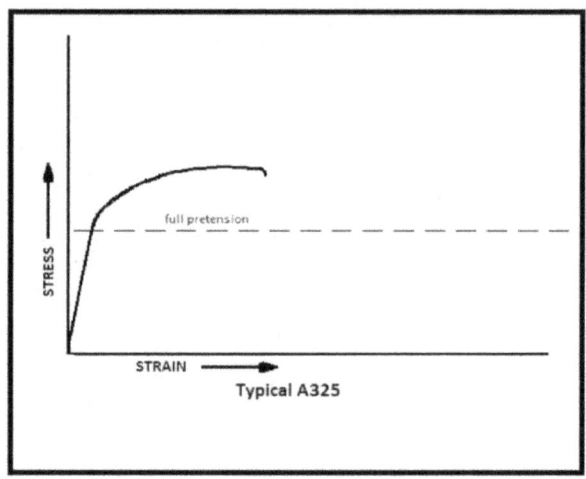

SOME COMMENTS ON "SLIP-CRITICAL"

There are times when some small movement (or slipping) of the plates in the plane of contact may be desired for some reason. One might want to allow for some thermal expansion or other slight shifting. One may like to hear the squeaking or groaning of a building because it reminds them of home. Or, someone may just not care (a very common reason.)

There are also situations where there is to be no movement of a joint - no slipping. Sometimes even 1/16 inch of movement -slip- in the plane of contact is not to be tolerated. Even 1/16 here and there in multiple joints can add up to something big. The design may have used oversized holes or slots for ease of erection, which, if allowed to slip could cause a significant movement that would become a usability issue. There could be vibration from machinery or other dynamic stress reversals that could cause fatigue in the bolts. Sometimes one might not want the full shear forces to bear directly on the bolts. There could also be effects from slipping at joints which could shift or redistribute forces at the joint in a bad way, or increase the effective length of members. So, they came up with Slip-Critical Joints!!!

Slip-critical joints are a type of pretensioned joint which are designed with frictional forces that resist shear in the plane of contact of the plates. This also means that **all** that has been said so far regarding pretensioned joints apply to these *plus* a few things that only apply to these.

As with all other joints, tension protects the bolts in a slip critical joint, and, as cannot be expressed enough, lack of tension is dangerous. It may even be more important for the protection of the bolts in a slip-critical joint and for the safety of anyone standing under it - even the people you don't like. Also, just because this type of joint involves some extra *calculated* resistances to shear, that does not mean that one could be careless in either fit-up of the joint or installation of the bolts since if those are not done right one cannot assume that there will be any frictional forces at all! Naturally, the use of these extra factors of shear resistance **only** resist shear and have nothing to do with resisting separation of a joint in a tensile manner. Therefore this discussion must center on shear resistance.

This concept of slip-critical joints is based on the concept of the *coefficient of friction*, which expresses friction forces in the plane of contact of any two things as a fraction of the compressive force normal to the surfaces in contact. It can only function as a joint designed as slip critical if that coefficient is .33 or greater (Class A surface.) Less than that still created some friction, but not enough to be relied upon to protect the joint in the manner prescribed for slip-critical joints. The shear resistance is the compressive force multiplied by the coefficient of friction at the area of contact. For example, a 3/4" A325 bolt pretensioned to 28 kips with a Class B surface coefficient of .5 will be capable of 14 kips of frictional slip resistance in the plane of the surface, assuming that this refers to a single isolated bolt in a correspondingly small annular area of contact around the bolt rather than the entire group of bolts, a single shear plane, standard holes, and, ignoring such factors as resistance factors or allowable stress and safety factors used in the formulas, as well as other pieces of reality. (Or, it would be 13 ½ kips when using LRFD strength limit state, also ignoring a few things.) That value of frictional shear resistance can be affected by the number of shear planes through the joint (since the friction forces are created at each of those planes of contact), or reduced by the use of other than standard holes, reaming or distortion of holes to fit the bolts in, or by various amounts depending on the type of nonstandard hole. The example used was for one bolt, but the total resistance of the joint depends first of all on the number of bolts in the joint and their arrangement in the joint, which determines the compressive force on the surfaces in contact.

The values of the slip resistance mentioned above are only the theoretical maximum. Although friction is not considered to be affected by area of contact under normal conditions, that can be different at extremely high pressures. Total slip resistance in a joint depends on things such as the area of steel in contact in the pressure affected area around the bolt, which involves the number of bolts creating friction. If the plates are not in contact at many points, no friction exists by those bolts. Bent or wavy plates or plates not aligned well may have much less slip resistance, if any, and the distribution of friction forces, if they exist, may be highly uneven, and if there is any possibility of movement of the plates (i.e. not in firm contact) then there will be no slip resistance. In the real world there are forces on joints attempting to pull the joints apart. If such is enough of a consideration to consider the joint under *combined shear and tension* (such as if prying action is allowed to occur at a shear connection for whatever reason) then the slip resistance of the joint is reduced. It can be reduced to nothing if the joint is not assembled or bolted properly or if the plates are out of alignment or the contractor got sloppy in some other way thinking that whatever effort they made is way more than enough and doesn't see the point of following the directions of the structural engineer or the researchers who developed the codes. The value of the theoretical slip resistance in the presence of tension loads on the joint (I don't mean *bolt pretension*), which **always** exist in some form even if neglected in design, is then multiplied by a reduction factor calculated by {1 minus [(tension produced by the loads on the joint) divided by (1.13 multiplied by the number of bolts in the joint and the bolt pretension)]}.

It can be seen by this that this amount will always be less than one and that when something like theoretical slip resistance or anything else is multiplied by a number less than one *it is reduced*. But, that reduction in slip resistance can be improved. If the number of bolts in the joint, or the tension in the bolts is increased (such as by using bigger bolts with a greater pretension value *or by simple following directions and pretensioning the bolts to the required value like contractors are supposed to do*), then the ratio being subtracted from one becomes smaller in relation to the factored loads designed to be resisted by the joint and the value of the reduction factor approaches one (one being theoretically no reduction in slip resistance.) Perhaps in simpler terms one would say that it should be obvious that if the compressive force on a joint is reduced for any cause, then the slip resistance is also reduced (less friction). Tightness therefore, as in all other cases, prevents joint movement *which prevents damage* rather than causing damage as is so commonly believed by both contractors and even some inspectors. That cannot be repeated enough.

If the compressive force (bolt tension) is not enough, then the bolts will be subject to greater shear forces, and at the same time, so will the connecting material. If there was not compressive force creating friction then the bolts would have to take the full shear load, and that may or may not be desirable depending on calculations and whether or not threads are included in the shear plane. Any time there is any bolt tension in a shear connection there will be some frictional shear resistance, however small, which will reduce the shear forces on a bolt, at least until the forces reach the point of causing the joint to slip into bearing on the bolts. Even though the theoretical ultimate shear strength of the bolt in itself is not considered to be greatly affected by pretension or lack thereof, pretension reduces the *shear stress* which must be resisted directly by bearing on the bolt if the joint is in danger of slipping. The frictional shear resistance of a joint can also protect the plates, and can mean the difference between a failure of net section through the holes, or resistance to failure by the entire gross section. So even if a joint is designed as snugtight tension still provides a measure of protection, even when tension is neglected in design on a simple shear connection.

Since the coefficient of friction is a property of the surface of the material, one should pay attention to what goes between the plies, and the conditions of the surface. The coefficient of friction for the class of surface will be the coefficient of friction of the coating and not the coefficient for steel on steel. Materials such as unqualified paints or zinc act like lubricants and can actually help the joint to slip. Creep can be made worse by coatings, which reduces the clamping force over time, and any reduction in clamping force will reduce the friction and shear resistance, since friction is created by pressure. The actual area of contact can matter, since if the plates are not in contact there is no slip resistance. If the plates are not in good alignment (<1/20 slope), then there will be areas of the plate not in adequate contact (like the whole thing), or enough tight contact to create friction and provide slip resistance. Bent or warped plates damaged from erection practices will have areas that are not in good contact. In such cases a joint may be within tolerances for other pretensioned joints (i.e. the connection might meet the definition of "firm contact" and there might not be a possibility of movement of the plates in line of the axis of the bolts considering tensile forces on the joint only) but such conditions reduce the amount of friction force produced between the plates around the bolts and therefore affects the slip resistance along the plane of the connection. Since bearing area can matter in the sense of the number of bolts contributing to the friction produced, oversized or distorted or slotted holes reduce slip resistance by removing contact area around the individual bolts, changing the pressure effects on the plate, and therefore <u>must not occur unless designed as such</u>.

As with other situations, it is the force *between* the plates that matter. The tension in the bolt is a means to an end. The compressive forces on the joint which resists the separation forces is a means to an end. With the slip-critical joints being discussed here **keeping the joint from moving is the end**. The greatest threat to the bolt remains being too loose, not, as some may imagine, being too tight. The factors of safety that may be designed into a connection are NEVER an excuse for sloppy work. Proper fit-up and bolt installation are critical to the performance of a slip-critical joint, as with all other joints, or the slip resistance may be not simply reduced by an amount that some contractors think is tolerable because of their fantasies regarding overdesign, but the resistance of the joint may actually be effectively <u>nonexistant</u>! There is never a tolerance on stupidity.

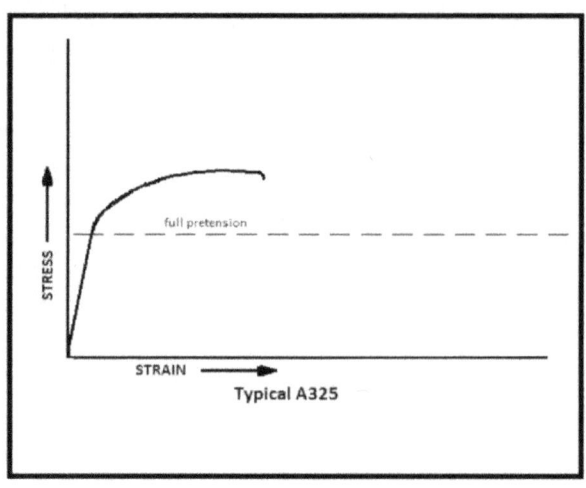

IS IT RIGHT?

PRETENSIONED JOINTS

It is assumed at this point in the discussion that the joints have been snugged properly -- meaning that <u>all</u> bolts in the joints have been snugged, and all joints have been fit with parts of the frame in proper alignment and plumb. It is assumed at this point in the discussion that surface conditions have been verified for joints designated as slip-critical, as well as others.

It is assumed at this point that washers were used everywhere they were required either by design or by code requirements. It is assumed that the bolts are the proper length for the joint or hardware installed so that the nut does not hit the thread runout, preventing the bolt from ever serving its purpose. It is assumed that the plans have been checked for any joints designated "X" meaning that the threads are to be excluded from the shear planes, the others being designated "N" meaning that the threads are allowed to be included in the shear planes.

Are all the bolts used in a single joint identical with no variation in length or length of threaded section included in the grip? If not, then the individual bolts might behave differently under some conditions and possibly not work as a team.

Is pretensioning required by either design or by code requirements (those two don't always match), or are the bolts merely erection bolts for a welded connection and are not part of design calculations, such as for some tube connections? Are there surface condition requirements- a coefficient of friction or surface class such as A, B or C? Are these bolts expected to share a load with welds? Is a certain type of high strength bolt specified, such as A325, A490, A449, F1852, F2280 or some proprietary bolt? Is there a pretensioning method specified? Have proper submittals been sent to the engineer regarding the contractor's intentions? What testing or sampling has been done? Are the mill test reports available for the bolts on arrival at the jobsite? Have all things been verified <u>before</u> proceeding?

Has reaming of holes been necessary? If so, were those modified holes which are now slots noted and marked for necessary corrective action such as plate washers if necessary? Has the structural engineer of record been consulted regarding any of the changes that are always necessary to make a building go together? Are there geometric matters which would make it difficult to access a bolt with either a TC gun or a 6 foot torque wrench? If so, another method or configuration must be submitted for approval, such as a welded joint or possible adjustment of some members.

Assuming that these matters have been reviewed <u>before</u> proceeding, then it is time to pretension the bolts where required.

Remember back at the beginning of the project *before* the steel had been erected and the bolts were still in <u>unopened</u> sealed containers straight from the supplier? There must have been an inspector present at that time for the Pre-Installation Verification required by code. If not, then it becomes much more difficult to pass a structure later and often involves reportable violations. The contractor should **never** proceed without the Pre-Installation Verification and I have never witnessed anything other than 100% failures when contractors do so. It becomes very painful for the contractor to sort out these problems later, when they could have been avoided by cooperating and following the rules. As indicated above, there is far more to bolting than simply shoving the bolt through a hole. Getting it right is important for safety. The level of safety of a bolt can be viewed in an over-simplified form as actual pretension minus required tension, which should always be a large positive number.

There is no way to look at a bolt and know how much tension exists *inside* it, unless it is obviously loose. There is also no way to look at a joint after it is fit and tensioned and know how much pressure and friction exist *between* the plates. All one can do is guess or estimate the possibility of success or failure based on the

Pre-Installation Verification of the procedure along with observation of the fitting and erection to eliminate variables affecting either the tensioning or the potential for relaxation or detensioning of bolts, such as many of the factors noted above and in the previous sections.

It has been shown in previous sections, and should be repeated, that all pretensioning is turn-of-nut, and that all pretensioning methods are actually methods of verification. This is because all methods involve observation of a click, a turn or a snap, which are referenced to the Pre-Installation testing.

Unless the nut is turned, no tension is produced. No tension is produced because there is no elongation of the bolt. No elongation occurs unless the nut is forced to advance along the threads, with the amount of elongation being related to the thread pitch. (The nut threads "pull" the bolt threads out of the joint.) The angle of rotation is achieved either indirectly and less reliably through a determined and <u>verified</u> torque applied to the nut and assumed to produce the necessary rotation, or through the more reliable method of measuring the actual angle of rotation of the nut in relation to the bolt. The angle of rotation is directly related to elongation through the thread pitch, and so the pretension is directly related to the angle of rotation also for a given bolt/nut lot. It could also be said that F1852 bolts and DTI washers also accomplish the same goals indirectly, but those methods are no different. One of those methods measures force applied (huge unpredictable spread of values) and the other method measures the net effect on the bolt (much more accurate as long as it is done correctly and all required procedures are followed and verified.)

There are advantages and disadvantages to each method, usually trading reliability with efficiency. Also to be considered with efficiency of a method is the number of times a contractor will have to rework each joint after the usual 100% failures (often even repeated 100% failures.)

Part of the Pre-Installation Verification which must take place <u>before</u> installation of the bolts is testing the method on a load-tester such as a Skidmore to verify that the method will produce at least the minimum level of pretension required by code, and better yet, exceed it enough that it will still meet the minimum code requirements after relaxation at the joint or shifting of the steel. Nothing in high strength bolting is prequalified. One cannot look at a chart and decide that they will use only that torque or rotation. That itself would be a code violation. Charts completely miss the point, as well as those who rely on them. The point is *the effect produced within the joint.* Charts also cause people to think that they don't need the *code mandated* Pre-Installation Verification. All methods must be tested and **proven** effective **before** proceeding.

On the day that the contractor decides to tell someone that they are proceeding (which is usually not the same time that they had actually proceeded), an inspector arrives for identification, sampling and testing of the *approved* methods, assuming that they intend to actually use a method that they received approval to use on that project. A load-tester is set up which measures the tension in the bolt by assuming that it is approximately equal to the pressure that is applied by the bolts on the plates. It measures the effect of the method on the joint under perfect conditions (which usually don't exist in a real joint.) The contractor will sometimes look at the gauge and think that it is reading in hundreds of foot-pounds when it is actually reading in *thousands of pounds of pressure or kips*.) Torque has no meaning to that load-tester, which is unable to measure torque and has never been intended to or needed to. Torque is not what anyone should be thinking about. Torque by itself has no meaning in the codes and cannot directly meet code requirements. Only tension can do that. The load-tester only measures the *effect* of the tightening and ignores how you achieved it. It looks to the end result, which is what compliance with the codes is all about. Therefore, the contractor and inspector must also be thinking of the end result or it will be difficult (and possibly even unlikely) to achieve that required result.

At that time, the contractor will think that it is the *bolts* being tested in the load tester. But that was what identification and sampling was for. The bolts will be tested in a lab when required. It is the method of pretensioning that is being tested. Even so, it is often not even the method being tested but the contractor's ability to perform it as required. It will demonstrate the contractor's understanding of the principles involved.

The nature of the load test ceremony is often lost on people. It represents what is <u>possible</u> under perfect conditions. The results only have relevance when the contractor accurately reproduces the test conditions in the actual structure and follows the procedure. Not following the verified procedure is a documentable code violation. It is documentable because correction will be necessary before the bolts will have any real chance of being at the right tension.

One of the less predictable variables is the snugging of the bolts. The initial tension can hardly be guessed. That is true for an individual bolt and also for groups of bolts. One may look at the gauge on the load tester and one may snug a bolt to any specific pressure, such as a percentage of the full pretension or proof load, with great accuracy and then qualify a rotation past that point. The problem is that at the joint there is no gauge to look at for an initial tension or initial rotation for snugness and the ironworker may simply guess that it is good enough, which usually greatly overestimates the actual tension. They always use greater care on the load tester to get it right than they actually use in the actual joint, but that doesn't mean that it is intentional. It is easy to make an effort once down on the floor and during a test when they don't have someone watching how many joints they get in a given amount of time, and much more fatiguing to do it thousands of times up in the air as quickly as possible. Also, a load tester tests only one bolt and does not demonstrate to the contractor the importance of proper snuggification and tensioning of the <u>entire group</u>. So, that means that rotation for pretension begins at a different point for each bolt in the joint. That also means that the joint has varying degrees of tightness and also looseness, which can cause relaxation of bolts even if the procedure was nominally followed for an individual bolt in the joint or F1852 bolts were snapped. So, that over-simplified margin (actual tension minus required tension) had better be a large positive number.

    Pre-Installation Verification therefore is far more than a formality. It is the time to try to eliminate potential problems. It is a time of demonstration and learning, and quite often of shock and surprise or the supposed expression of such. People can often miss the point. I have seen a situation involving establishing a torque for A490 bolts with the foreman watching two people hanging from the end of a six-foot torque wrench to get it to the right gauge pressure on the load tester and then hand the ironworker who would be doing the tensioning an 18 inch ratchet to use. That was done repeatedly, and only after a week of 100% failures did the concept begin to sink in with the foreman that what was done on the verification test was the **mandatory** procedure at the joint if it is to pass. It represents the <u>minimum</u> effort necessary to get it right, and no less than that can be accepted.

    There is always an inexcusable display of shock and disbelief when ironworkers begin using non-tension controlled bolts after using only F1852 bolts which are so common. The problem comes from the way a TC gun applies the torque. The torque is balanced: force is applied to the nut and an opposing force is applied to the splines in the opposite sense. One does not feel the force. When one uses a "calibrated wrench", including impact wrenches, one must fight the opposing force one self. To push one way (on the bolt) it is necessary to push in the opposite direction on something (or someone) else. One pushes on the bolt and the bolt pushes back. Big bolts require multipliers which must also push and be pushed. For A307 bolts, one's arm is enough to push with. For high strength bolts the multiplier on the torque guns must push off something in order to apply the torque to the bolt. The process is naturally unbalanced. It always seems to those contractors like they are being required to make those bolts much tighter than the ones that they claim to have been snapping off for however many decades as they say. The reality is that the "tension control" F1852 bolts are often the ones which must be even tighter, since they have such a HUGE spread in tension values that the manufacturers must make them to overshoot the minimum a certain amount or the average value will fail. The difference is that they have to feel the actual force now when they were completely oblivious to it before when the force was even greater!

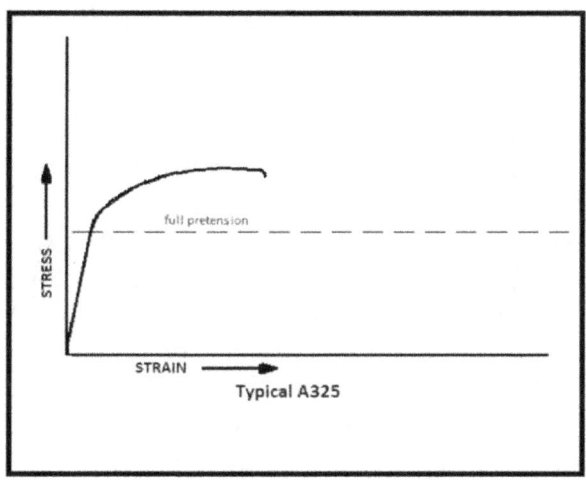

IS IT RIGHT?

Pretensioned Joints: effect oriented methods

Turn-of-nut is an example of an effect oriented method, but all methods are really turn-of-nut. All methods apply force (torque) to either the nut (preferred) or the bolt (acceptable when necessary) to cause rotation. The difference between effect oriented methods such as name brand turn-of-nut and the torque oriented methods is that this one focuses on the <u>effect</u> of the force on the bolt rather than the amount of force applied. This is considered much more reliable, as long as it is done right. The reliability factor with these methods theoretically (but not calculated as such) jumps from 1.13 to a much higher value which can't be used in design unless proven.

So, why not always use this method? It gives the appearance of being slow and labor intensive, especially when compared to snapping TC bolts. Appearances can be deceptive.

When someone is using torque oriented methods, such as a "calibrated wrench", they simply do it until it clicks, snaps off or stops turning at the desired torque value. They erroneously also think that this method is faster than turn-of-nut, even though the nut must still be turned the same amount if it is done right. In reality, it doesn't take any more time to watch rotation. The difference that they perceive in time and effort is only an illusion caused by the fact that turn-of-nut tensioning doesn't allow them to give up at a certain effort (torque), and they have to keep turning until it is right. Theoretically, there should be no difference in effort or rotation between the two methods if either is done right following the procedure qualified in the Mandatory Pre-Installation Verification testing, but there are other factors you can't see on a load tester. The ironworker does not so much need to observe the bolt rotation during "calibrated wrench" tensioning, although that can still be done by observing the rotation of the socket. With turn-of-nut, one must observe and measure the rotation. This can be done by marking the bolt/nut or even the socket (assuming that there is no rotation of the bolt assembly).

So, why is this more reliable, especially with matchmarking? It goes back to the stress/strain curve as all things do, and because it measures the actual effect on the bolt where torque oriented methods ignore the effect or make assumptions only. A given degree of rotation past snug <u>always</u> produces a given degree of elongation for a particular thread pitch and length for an individual bolt/nut lot, within manufacturing tolerances. That can't change, even if the bolt loses lubrication, because it is part of the machined geometry of the bolt.

The initially unknown factors are how far a nut must rotate (bolt elongation) to produce the required pressure between the plates (pretension) and how much force it may take to produce that pressure. There are geometric factors involved such as the length of the threaded section within the grip, such as fully threaded and partially threaded, and the total length of the bolt within the plies (grip). The rotation (elongation) necessary is to some extent related to the hardness of the bolt (yield point) and the stress/strain curve, since pretension should be assumed to just about reach or cross the yield point, but might not. Location of that yield point makes a noticeable difference when comparing A325 bolts to A490 bolts. How much force is required depends upon all factors noted above which affect torque, but to a certain degree that is being ignored, except for establishing an arbitration torque. Of course, the amount of torque required is important when one is dealing with delubricated or slightly rusting bolts or dirty bolts or thread damage or other matters which dequalify a bolt, since there is an increased risk of damaging the bolt before producing the required pressure (pretension). But, one typically need not worry about residual torsional stresses since: (1) they shouldn't be using bad bolts, (2) the residual torsional stresses are said to have little effect at the point of plate separation and therefore are not to be considered additive with tension, and (3) because it is just like using a torque wrench as far as the applied force is concerned.

It must be repeated that the tests performed on a load tester represent perfect conditions which may not exist in a real joint. It only tests one bolt at a time and can't check the tightening sequence the contractor will use. Real joints may have compressible coatings such as paint or zinc or worse things which must be removed. Those may require a greater rotation to counter the effects of the significant relaxation which can occur in a short time.

Each bolt diameter will have a different rotation. Even within a given lot there can be a spread in rotations and individual bolts will have a noticeable individual personality, especially if the yield point is crossed. It can be a big difference, such as between ½" bolts and 1" bolts, or it can be barely noticeable. This is to a great degree because of the different thread pitches used with different diameters, but from one lot to another there are differences from manufacturing tolerances and methods and metallurgical factors. An A490 bolt will also have a greater required rotation than an A325, even though the thread pitch is the same for a given diameter, not just because the steel is harder (they have the same modulus of elasticity and behave the same in the elastic range) but because they require a higher level of pretension. That higher level must be produced through more rotation.

One can't simply look at a chart and think that no testing is required. Testing is always required. Each bolt will be a little more or a little less than others. Different bolts may have different ratios of threaded to unthreaded sections within the grip, such as fully threaded and partially threaded bolts, and more (some people say all) of the elongation is concentrated in the threaded section because of the smaller net tensile area, especially if the yield point is crossed..

The efficacy and accuracy of an effect oriented method such as turn-of-nut, like all methods, is affected by such factors as joint fit-up, proper snugging of *all* bolts in a joint and proper tightening sequence. A given rotation past snug is meaningless unless there is a predictable snugness to start from. If a joint moves or settles during or after tightening then the pretension in any bolt tightened before the change could drop to zero. A sequence of tightening which causes the bolts to relax as one progresses means that even matchmarking can't be relied upon as an indication of actual pretension. It only implied that a bolt was turned at one time and means nothing more.

This method (turn-of-nut) is only reliable under the right conditions, as with all methods. This method is also perhaps the most sensitive to variations in snugness. But when done right, it can be very reliable since you can witness the actual elongation of the bolt, which, if begun from a reliable starting point, always relates to a particular level of tension, as shown by the stress/strain curve.

At first this method may appear to be slower, especially when matchmarking is used. Reality is a little different. Yes, it means that they are paying attention to the rotation. But, all methods are turn-of-nut. Torque oriented methods, if done right, produce exactly the same rotation, although less reliably. Therefore the amount of work is the same. The difference is that with torque oriented methods they are ignoring the actual rotation. They may feel more frustrated with turn-of-nut because they can see for themselves that they are not getting it right, whereas with a torque wrench they can't see the nothing they are actually accomplishing with the amount of effort they thought was good enough. The rotation is still the same. If rotation is the same, then work is the same since in terms of physics work is the applied force multiplied by the extent of the effect such as how far the nut turned. One method is more accurate than the other. They may speed through with one method and think they are saving time by using some perceived shortcut, but then after the common 100% failures certainty begins to look more affordable.

Because they are now paying attention to rotation, they will make the same idiotic display of bogus surprise when they are told (and shown repeatedly on a load tester) how much rotation is required. When they see it they seem to miss the point that it is pressure (tension) that is required. Any method which does not produce the *code required minimum tension* (pressure) is wrong and is a code violation. Torque, or effort, is irrelevant. If it does not produce the required pressure on the load tester (which represents perfect conditions), then it will be even worse in an actual joint. The pressure on the gauge (kips not foot-pounds) will relate to a particular elongation of the bolt per the stress/strain curve and the elastic modulus (which is also a part of the stress/strain curve). That elongation relates to a particular rotation of the nut because of the thread geometry, but will vary from lot to lot. That applies to all methods equally. All methods use the exact same stress/strain curve and the exact same modulus of elasticity because those are characteristics of the bolt and are unrelated to how the nut is turned. It therefore will take the exact same effort for a particular bolt no matter how you look at it. Where were they back in the days when they turned all bolts with a far greater rotation by default on every job? Those bolts didn't break, and those joints could be much more reliable than the garbage thrown together these days after someone decided that the bolts only have to be just tight enough, and lowered the requirements to something they thought people might actually do.

When they look at the effect (rotation/elongation) they will claim that it is more than required and that even 1/3 turn past snug will break the bolt. This is just another case of the same whining and display of incompetence seen with any method other than snap-off TC bolts (since with those they remain completely oblivious to what is actually taking place). The fact that one can look on a mill test report (from some manufacturers) and see that the bolts were tested by rotation a few *times* further, such as a full turn or more for longer bolts without any damage, seems completely lost on people, including inspectors.

As has been noted, just because 1/3 turn (or whatever actual rotation is required for a particular bolt) corresponds to 70% of the tensile strength of the bolt, that does not mean that 2/3 turn will produce 140% of the tensile strength. As with all things in bolting, one looks to the stress/strain curve for the facts. At the 70% tensile strength level the bolts theoretically reach the *minimum* required yield point. That is <u>never</u> a bad thing. It only means that the stress/strain curve becomes *nonlinear*. That is not a bad thing either. What happens in that range (referred to as the plastic range) is that a given rotation (elongation) corresponds to a smaller increase in tension. Stated another way, a smaller increase in tension corresponds to a larger rotation (elongation). So, one can go way past the minimum rotation without approaching the point of doing damage. That is why the bolts are tested for such a large rotational capacity far in excess of what will be required on a project and are expected to show no damage.

This also means that doing less than required has a greater impact on the actual level of pretension than doing more. This is because the bolt will still be in the elastic (linear) range where rotation (elongation) corresponds to a greater difference in tension, and a greater change in tension is related to a smaller change in elongation (rotation).

Let's say that a bolt is rotated a certain amount which is a certain number of degrees short of the required minimum and another bolt is rotated an equal number of degrees past what is required. The bolt that didn't go far enough may be 10 kips low, for instance, which is dangerous to the joint and everyone who depends on it and dramatically increases the chance of bolts breaking, since the bolts not fully pretensioned will not take the load from the bolts that were. They can't share the load equally. The one that was rotated an equal amount past the minimum may only be 3 or 4 kips past the minimum, depending on the bolt, which does nothing to damage the bolt and is nowhere close to the limit. Such a bolt is less likely to break because it has a greater capacity to transfer force through the joint without the joint separating, and the tension in that bolt is close enough to the other bolts for load sharing among the bolts. It is therefore safe to go further – very safe – but dangerous to not go far enough, as with all things in bolting.

Direct tension indicating washers are also in the effect oriented category. It takes a specific pressure to crush the bumps or squirt the paint. It is therefore certain that the required pressure (clamping force) was created by the bolt on the joint at one time.

Under the right conditions they can possibly be considered more reliable than turn-of-nut measurement. This is because with turn-of-nut measuring, as noted above, there can be factors which can cause the rotation to be deceptive and give an impression of a pressure which may not exist when it is done wrong. A load indicating washer proves that the pressure was in fact created on the joint (not necessarily in the joint).

As with all methods, this method is also turn-of-nut, since without rotation there can be no tension. But with this method one is looking beyond the angle of rotation and observing the effect of the elongation, at least at the surface of the plates. Similar to "calibrated" wrench or snap-off bolts, or even turn-of-nut, one turns it until it is right.

As an effect oriented method, this also ignores how much torque is applied to make it happen. This may help to mitigate some of the effects of the usual slop and carelessness. It also helps to address the issue of trying to match the perfect conditions of the Pre-Installation testing with real joints.

Pressure on the plates occurs because of the tension in the bolt. Tension occurs in the bolt along with its close friend elongation, which have a linear relationship up to the yield point, which is near or at the full pretension level. Elongation occurs because of rotation and is a geometric issue.

So why not always use such a reliable method (if any method could be referred to in such a way)? Cost. It takes longer than any other method mentioned in this work since the gaps must be measured or the silicone paint observed (faster). The washers cost and increase the material cost of the project.

Is it guaranteed to be right? Is one always certain that the joint is good? No such thing is ever possible. Engineers come up with new gadgets to deal with one issue that had been annoying people, but that can't fix everything, since it is not the engineers who are using or inspecting them.

A washer of this sort is a load tester. They can even be used for qualifying a procedure, especially with the short bolts on every project. They also, like a Skidmore, only test one bolt at a time and do not test an entire joint, even if all the bolts in a joint have them. They can only tell you that the pressure was right at some time in the past. They don't necessarily tell you what the pressure is right now – especially with the squirter types. All bolts are tensioned one at a time. That can change things as one progresses from one bolt to another. They can't tell you if the tightening sequence is correct.

As with all methods in bolting, pretensioned or other, the steel must fit together. If the plates can't be brought into contact because of bent plates or other reasons, then there may be little pressure inside the joint, especially if something shifts later, and there would be an increased risk of relaxation and the plates eventually yield and creep. There would certainly be very little slip resistance. All one can measure is the pressure exerted <u>outside</u> the joint, and only guess that it is being transferred to the inside surface of the plates.

There are also other methods, such as expensive specialty bolts and ultrasonic elongation measurement, but those are far less common. The ultrasonic method is perhaps the most reliable, since it would detect relaxation of the bolt if there was a shift in the plates or the wrong tightening sequence was used, since any loss of pretension would be related to a corresponding decrease in elongation.

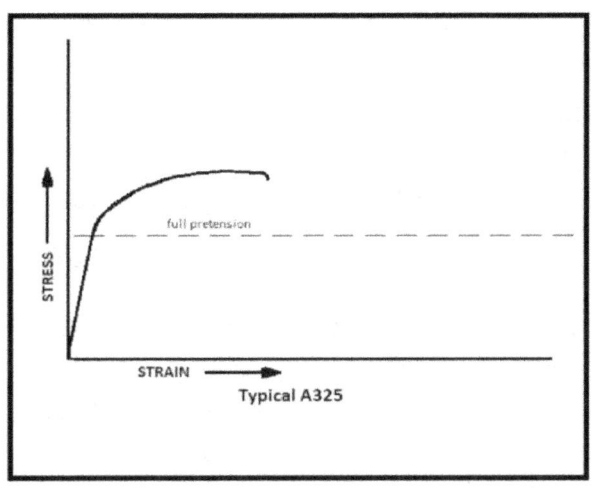

IS IT RIGHT?

PRETENSIONED JOINTS: **TORQUE ORIENTED METHODS**

And what about those mysterious F1852 bolts? They were invented to make stupid people think they are ironworkers. There is no nice way to say it, since that is the result. Stick the gun on and pull the trigger – that is all people think there is to it. People think that it is so easy that they can neglect <u>every</u> rule. Yet, those bolts have even more rules. Using those bolts never excuses negligence.

One must keep in mind that snapping off bolts is never the point. All one sees when one looks at it is a snapped spline section. Unless it creates the required tension in the bolt, and the bolt maintains it, it is meaningless. Unless that tension creates the clamping force on the joint it is meaningless. Unless that clamping force creates the required pressure *between* the plates and the necessary friction for slip resistance when required it is meaningless. If there is relaxation and loosening of the joint as one progresses snapping them off in the usual happy-go-lucky way it is worthless to deadly.

The ease of snapping them off encourages carelessness and encourages unqualified contractors to avoid learning the rules. Snapping off bolts seems so easy (which it is) that people forget all the other matters of bolt installation which come first. Proper erection and fitting of bolts matter just as much. Proper snugging and **all** other considerations noted in every section of this work apply to F1852 bolts.

When one of those bolts is snapped, that *only* means that a certain torque has been applied to the splined section, which is always *outside* of the joint. That's all! It does not mean anything more than that. It does not automatically imply any effect *within* the joint. Anything already stated which affects torque reduces the effectiveness of F1852 bolts just as much. If all the force is used to overcome friction from any source (under even good conditions 90% or more of the torque is used to overcome friction only) that reduces the value of the torque going toward tensioning the bolt. If snapping a bolt only brings the plates into firm contact, the effective tension will be zero.

Pre-Installation Verification is necessary to determine that snapping those bolts is enough to produce the code required pressure. People feel that by making something simple it has to be so certain that verification is unnecessary. 1" F1852 bolts are just as soft as their ordinary A325 counterparts. Bolts are often not taken care of properly and I have heard many times the line "I've never seen one of those fail in all the year(s) I've been in this business", whatever that means. Well, I have, and so have many others, probably including those same contractors if they really had as much experience as they claim. Nothing is ever prequalified in bolting. (Contractors even say things like, "the building(s) I worked on were done right because they didn't fall down in the 1994 Northridge Earthquake in California, and they will always be just fine – in Wisconsin.") Yes, I actually do hear things that stupid!

F1852 bolts are similar to any other method relying on torque, such as calibrated wrench pretensioning. F1852 bolts are called "tension control", but they are really "Friction Control" and have nothing to do with tension. They only call them "tension control bolts" to make them sound like they meet code requirements. The principle of the torque method is applying a certain amount of force, torque in this case, and making the assumption that it will turn the nut an amount, producing elongation and therefore tension and pressure of an amount that can *only* be determined by testing during the Pre-Installation Verification which is *mandatory*. (With any method, the force involved is always torque, and the different methods relate to what one is focusing on –verification through either snapped splines, a click of a wrench, rotation of a nut, or crushing of bumps on a DTI washer. By "torque oriented method" I mean the methods where one focuses only on the applied torque without observing the actual effect on the bolt.)

Any torque oriented method is critically affected by anything which affects friction on the bolt. Proper storage of the bolts is critical. Dirt and rust will change the rotation (equivalent to tension) produced from a certain torque. The state of lubrication can also change before there is any noticeable rust. F1852 bolts could be induced to snap off before the required tension is reached, since they are "torque controlled" and will snap off at a spread of torque values and don't care if *any* tension or rotation occurs. Any change in the lubrication of F1852 bolts disqualifies them, unless the manufacturer gives detailed instructions on re-lubrication which must be followed exactly. It's easier and cheaper just to take another bolt out of the can, and the bolt should be considered disqualified since no one is really going to go through the trouble of that and requalification. Bolts that have been on a project for some time may have met the requirements at one time, but that doesn't mean that they will still produce the same tension later. Under perfect conditions 90% or more of the torque applied to the bolt is used to overcome friction. Very little of the force actually goes toward producing elongation, and that can very easily be made worse. It only takes a tiny change in the coefficient of friction, say 1% or 2%, to have a noticeable change in the tension produced. Therefore, if one ignores effect and focuses only on torque, conditions can easily cause that torque to have no effect at all.

As already stated, to call an F1852 or F2280 bolt "Tension Control" is very misleading. They are made to snap off within a particular range of torque values. That has nothing to do with tension directly, either the creation of the tension or the retention of the tension. It may be assumed that as tension in the bolt increases, the friction force at the surface of the threads increases requiring greater torque, since the friction force is the force of the pressure normal to the surface multiplied by the coefficient of friction. That means that the friction will be a proportion of the pressure on the surface of the threads with a small amount of trigonometry involved. That friction force has a direction opposite the direction of advance along the threads and opposes the advance. That pressure on the threads then becomes a shear force on the base of the threads. The force on the threads is distributed unevenly, with the first thread taking a larger share and the pressure decreasing with each thread after the first. That shear force shared between the threads and the surface of the shank is transformed into tension as the metal is pulled by that force. The nut will not slide along the surface of the threads unless the applied force is greater than the resisting force. If a certain maximum torque value (or range of potential values in the case of F1852 snap-off bolts) is used, a certain equilibrium will eventually be reached at a certain point where there will be no more turning. As the nut has been advancing along the threads to that point, the friction force at that point has increased to the point of being equal to the torsional strength of the spline section. A TC gun has been applying more force to the splines as resistance has been increasing, and hopefully the splines will snap before the bolt does. How much tension was created at that point? An F1852 bolt doesn't care. A torque wrench doesn't care. The only thing they know is that a torque value was reached and they have been designed give up to go no further than that regardless of whether it is right. With torque oriented methods, either splines snap or a wrench clicks or quits turning. Nothing more.

    Required torque can be increased, or decreased (much less common). The coefficient of friction can increase. Any increase in the coefficient of friction of the threads increases the resisting force produced at a given rotation thereby requiring more torque to produce a given degree of further rotation. The torque necessary to produce the required rotation may become excessive and especially beyond the capability of an F1852 bolt. That reduces the tension produced by a given amount of torque.

Look at the snapped off bolt. What do you see? If you see anything other than a snapped off bolt you're hallucinating. All it is a snapped off bolt. It only means that at some point in the past a torque was applied sufficient to break the splined end off. Anything else is pure conjecture, if not outright fantasy. You can't see the tension in the bolt. You can't see the pressure between the plates. One can only speak of a *possibility* that code required tension is present.

Look at the snapped off joint. What do you see? If you see anything other than a snapped off joint you're hallucinating. The only thing you can say for certain is that at some point in the past, or often at a discontinuous series of points in the past for each of the individual bolts in that joint, each bolt individually and separately had a torque applied sufficient to break the splined end. It means nothing more, especially considering that a TC gun can only snap one bolt at a time and can't snap off an entire joint at once, as fast as they may be. There is only a level of probability that it was done right, and that level of probability can easily drop to zero.

Done right?! But aren't the F1852 bolts designed to take care of themselves and are automatically right with no further consideration? Don't you only have to shove them in a hole (even beating them with hammers to get them in!) and stick the TC gun on them and pull the trigger? Aren't those bolts so good that you don't have to worry about snugging them properly or aligning and fitting the joint properly? (Yes, it gets that stupid.) Aren't those bolts so good that you don't have to pay any attention to any pattern of snapping them and there is no possibility of the previous bolt loosening ever? Once they are snapped, aren't they *always* tight enough and *never* have to be thought of anymore?

Every one of those statements is so common that one can conclude that F1852 snap-off bolts were invented to make stupid people think they are ironworkers. That does not mean that there are no benefits to their use. When done right and when they are treated like every other bolt, and even more carefully, then they can be good enough. The other bolts aren't necessarily going to be used any more correctly. Perhaps in actuality these bolts are slightly more reliable than the "calibrated" wrench method because F1852 bolts in theory snap off at a higher level of tension than required to account for the high variability, and the "calibrated" wrench method only applies the bare minimum, even though both methods may be the same in theory. At least one *does know* that a particular torque value *was* actually reached, and with the "calibrated" wrench method one does not know without observation of the torqueing or torque testing the bolts at the arbitration torque. But how good those snapped off bolts are depends on conditions and the overall circumstances of their use. Like everything in bolting, it depends on what took place before as well as during. It is easy to have them fail. It is easy to have a loose joint. All one has to do is not pay attention and get sloppy.

Proper fitting and snugging of the joint first is just as critical to performance of the joint as it is with any other type of bolt. Any gap in a joint can create potential for movement as one progresses. Even with an F1852 bolt the torque can be wasted closing a joint which should have been fit better before snapping the bolts. A snapped bolt only implies that a certain torque sufficient to snap it was applied. If a joint is capable of movement, then the next bolt snapped will cause the previous bolt to loosen, just like with all bolts. Snapping off the bolts in the wrong pattern loosens previous bolts. All you can see is that torque was applied to the splines. That can happen at <u>any</u> level of pretension. It can happen at an even lower tension if one goes directly from loose to snapped without snugging the entire joint first.

How long have the bolts been sitting loose in a joint before they were snapped? How much rain and dust or even paint or worse have the bolts been exposed to? They may have passed in the load tester half a year ago, but they are not the same bolts anymore, and would need to be retested.

Did some of those F1852 bolts get dirty? Throw them away. You can't clean them without an approved procedure from the manufacturer or use them, and it is not worth the effort. Are they no longer properly lubricated? Only the manufacturer can lubricate them, or provide the procedure for use of the exact same lubricant and application, because any change in lubrication affects the Coefficient of Friction of the threads.

Did you accidentally drop and lose the nuts which came in the *Fastener Assembly*, which the bolt, nut, and washer is referred to?

One can't just go find any nut with any lot number from any other can of the same size. Only the manufacturer can assemble them and they must be used as assembled by the manufacturer or they are no longer qualified – assuming that they were identified and qualified during the Pre-Installation Verification.

Do you need to remove some that were already snapped? Throw them away. You can't reuse those like you can ordinary plain nongalvanized A325 bolts. You can't resnap them.

The seemingly simple can have many considerations. In the case of the F1852 torque control snap-off bolts, there are even more considerations than other bolts. Yet when all above has been addressed, they can be a fast and efficient method which can be worth the extra material cost.

Much that was discussed above applies to *any* torque oriented method such as the "calibrated" wrench method, although there are slightly fewer rules than apply to other bolt types than with F1852 or F2280 bolts. Like every method, there are also some unique requirements.

Wrenches come in a few forms, such as electric, pneumatic, hand turned, working and nonworking. Some may be more *consistently* accurate than others. All should be calibrated for accuracy and all require Pre-Installation Verification. But, what is it calibrated for? Torque, not tension. The wrench knows nothing about the tension in the bolt.

Even though the principles are identical to the F1852 and F2280 bolts, one problem is that contractors using the "calibrated" wrench method will only use the barest minimum effort required. An F1852 bolt snaps off when it snaps off, not sooner and certainly not later. There is no choice of how much force to apply with those. For the "calibrated" wrench method the reliability factor of 1.13 seems almost too high, since that is only possible under well controlled conditions where the rules are followed.

As with F1852 bolts, the effect is being ignored. One is paying attention only to the force applied. The codes are only concerned with the effect. The current codes do not specify a torque value. The joints are not designed by torque. Torque does not hold the joint together. Torque does not resist loads on the joint or transfer force between the plates. Torque does not keep people alive.

Torque has absolutely no meaning unless it is proven with proper testing that it can be related to a degree of rotation of the nut or bolt. Rotation is the effect which the joints are designed by, since rotation is related directly to elongation, which is related to a specific tension, as seen in the stress/strain curve and demonstrated in the Pre-Installation Verification testing.

One could consider "calibrated" wrench pretensioning less reliable than F1852 and F2280 bolts because F1852 bolts will snap off at a higher torque than required to produce the required tension. F1852 bolts are what they are, at least when they first come out of the can. You don't snap them off later or earlier than they snap. Ordinary bolts aren't that way. They are just as tight or loose as one wants them to be.

One of the problems with this is what constitutes a "calibrated" wrench. On a normal project we are normally not talking about something that can be locked at a specific torque value, or even one which can be read closer than plus or minus 50 foot-pounds or worse. Contractors are used to pulling a trigger. Such automatic tools are grossly unreliable as far as a particular torque value is concerned. They are strongly affected by the power source. The contractors will change the settings or sockets or other configurations very often. Just because a torque value worked on a load tester, that doesn't mean that it will work in a joint or that the contractor is actually using that torque with any accuracy.

The contractors will point to the calibration sticker on the torque gun. Then you can show them on a load tester, repeatedly, that what they claim is the same setting is producing a 20% or more spread in actual pretension. And that was under perfect conditions which don't exist in a real joint!

Having a sticker on a torque gun – they should be calibrated often – has no meaning in a structure. It does not eliminate the requirement for Pre-Installation Verification. They only calibrated the torque gun for torque when they put that sticker on, not tension, since that would be impossible and they were never designed for that. The lab calibration of the wrench has nothing to do with a project, only the wrench, and not how the wrench is used. And, no matter how much they whine about how accurate the piece of junk is because it is clean and shiny, it needs to be verified continually.

Just because a setting worked yesterday, it is not the same setting after rattling in the back of a truck. Does the temperature stay exactly the same all day? It is not the same setting no matter how similar the dial looks.

And what is being verified in the Pre-Installation Verification? That it is producing the required rotation. It is common for settings on those guns to need continual adjustment. Did that thing produce a particular torque on the date on that sticker? So what? Torque has no meaning, even with torque oriented methods. That mistake is related to the belief that you are allowed to use some chart to eliminate testing. If it does not produce the required pressure between the plates then it doesn't matter what that dial is set to.

Manual torque wrenches which can be locked at a value are perhaps more certain, and usually have a much finer range of adjustment, but that implies that it is used right. It also takes more time, which always makes it less likely to be used right.

Like the F1852 bolts, "calibrated" wrench pretensioning is "friction controlled". Everything already noted above regarding torque and friction applies here and need not be repeated, except that what makes F1852 bolts less reliable makes "calibrated" wrench methods even worse. Even Worse! That is because F1852 bolts are designed to go past the minimum, with the possible exception of some of the 1 inch ones, which will help mitigate some of the factors to some degree. With "calibrated" wrench torque/friction based methods, the contractor will <u>never</u> be trying to do any more than the minimum. You might even call this method "effort control". There is no margin. That makes the usual factors affecting pretension worse for this method, since there is nothing extra to make up for the real world variables.

Remember, torque exists outside of the joint at the wrench. It does not exist inside the joint, except to the degree that it is converted to tension through rotation. We are only concerned with what goes on inside the joint. Only what goes on inside – tension and pressure – meets the requirements. Even one million foot pounds on the surface of the nut would be meaningless unless it is accompanied by the rotation which produces the elongation which clamps the plates together in a useful and real way.

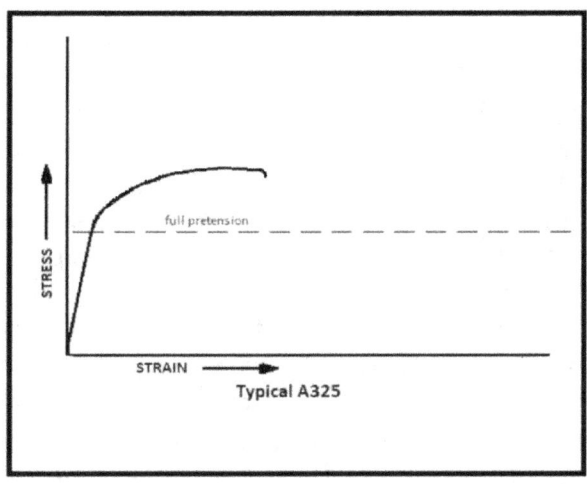

PRE-INSTALLATION VERIFICATION

As the name makes perfectly clear and cannot be excusably misunderstood, Pre-Installation Verification occurs **BEFORE** erection. It occurs before **ANY** bolts are installed. The bolts must still be in unopened cans with proper labels with all required information from the manufacturer and supplier.

The seals on the unopened cans are to be broken only when the inspector is present to observe. The inspector will then *randomly* take samples from each lot number, which is also to occur *after* having received the mill test reports for each and every one. The inspector will verify that the bolts, nuts and washers or assemblies of such match the label, including all things such as size, length and manufacturer's mark.

If anything is different, then it is a different bolt. A contractor may pull out an A325 and say that it is the same bolt because it is the same diameter, but is a couple inches longer – yes, it gets that dumb. If anything is different, including everything and anything, like especially obvious things like length, then it is a different bolt and must be tested separately.

Sometimes bolts may arrive with sets of three in plastic bags, specially chosen *by the supplier* for testing by the inspector. It should be obvious that such things are not *random sampling*. One should not necessarily even consider such things even identified no matter how similar they appear, because the fact that they were deliberately separated and specially handled means that they could have come from anywhere, and lot integrity must be maintained from the time of manufacture to the time of installation.

One set of sample bolts are sent to a laboratory for testing, which is the true material identification. Therefore, what is done in the field on a load tester is not being performed to qualify the bolt. It is obviously not that, and it is obvious that it is to qualify the contractor's procedure which should have been previously submitted to the engineer and approved for the project (or demonstrate to the contractor the need for a procedure).

Another thing which makes it perfectly clear that Pre-Installation Verification occurs before erection is the fact that one is not required to pay attention to which bolts from which cans are used where. That is because the RCSC Specification assumes that all testing has been done already on all bolts present on the jobsite and so all bolts used are the ones which passed and no other bolts will be used. So it is obvious that Pre-Installation Verification must occur before even opening any cans, and any cans found opened when the inspector arrives must be considered adulterated. Those bolts in an opened can did not necessarily come from that can and can't be trusted, especially with the tendency of contractors to combine cans which are close to empty so they can clear them out. Any bolts already installed did not necessarily come from any of the cans on the jobsite. It does not matter if the contractor claims that the bolts were supposedly already tested on some project a year before or more. You don't know what those bolts are or where they came from and you certainly can't assume that they had ever been tested since most projects are inadequately or incompetently inspected, if inspected at all. Many times an engineer on a minor structure doesn't really care and does not specify inspection in a project manual or in the structural notes. In that case, there is often no inspection of bolting at all, no matter what the codes say. Also, bolts that old from opened cans, even a few months, are not the same bolts anymore, and even if previous testing is somehow proven – for some other project – that does not qualify them for *this* project. Another project may have had material identification only, and those bolts may not have ever been placed in a load tester with a pretensioning method. **PRE-INSTALLATION VERIFICATION IS MANDATORY AND CANNOT BE SKIPPED.** It is **MANDATORY** that it occurs **BEFORE** erection begins, except that columns may be placed. Any bolts installed prior to this should be considered unidentified since there is not enough information marked on a bolt head to match it to a mill test report, and there is a tendency for contractors to use old leftover bolts which are not from unopened identifiable cans and have been rattling around in the back of a truck rusting for unknown years, if an inspector is not present to stop that nonsense. They will even use the wrong grade of nut if they can get it on the bolt, and a bolt assembly is only as strong as its weakest part.

When the **Pre-Installation** Verification is performed, once again, it is the method being tested, not the bolt. It is even more critical with methods such as "calibrated" wrench because it depends even more on the contractor. All the issues which can occur with F1852 bolts and which affect friction can occur with any bolt. Why is this friction an important factor? Isn't all you need to do is crank the wrench a couple hundred foot-pounds and it's good? No.

Friction is a factor because of the use of methods which focus on torque rather than the effect on the bolt. Anything which would affect the friction forces on a bolt will affect the amount of torque needed to turn it. This can't be repeated enough, because it is so commonly believed, even among inspectors who should know better, that torque is the whole point of this exercise. The greater the friction, the lower will be the level of tension in a bolt for a given torque value. It has no meaning if it doesn't produce the required pressure between the plates, as with any and all methods.

A seriously disregarded or disrespected part of this testing for "calibrated" wrench pretensioning on the load tester is snugging, since the real effect occurs in the joint where you don't see it on a gauge. Because Pre-Installation Verification is performed to qualify the procedure, it must begin at snugtight. That applies to any method. The snugtight condition, when done properly, provides a starting point for all else.

When they are using a "calibrated" wrench on a load tester, for an individual bolt snugging may appear less important because it has little influence on the final torque value and one keeps turning until the gauge reaches the required pressure. But, it is not the bolt being tested. Snugging is part of the procedure and must be performed in the joint before pretensioning. This, if done right, is necessary to prevent or reduce relaxation and loosening of the bolts as pretensioning progresses at the joint. Contractors should therefore pay careful attention to it since it will affect whether the joint passes later verification testing.

Therefore, one should always keep in mind that it is the joint that matters most. When testing in a load tester one must remain aware that it can only test one bolt at a time, and even then it is only under perfect conditions, and it does not test the whole joint. It only establishes a procedure which must be followed. One may not see the effects of snugging with "calibrated" wrench testing on one bolt in a load tester and it remains invisible to the contractor at the joint. It becomes apparent how critical it is when the joints fail testing, which is normal, or when light can be seen between the plates that aren't being held together well, which is also normal. One cannot forget this limitation of the Pre-Installation Verification, which is only a limitation if one forgets that it is the <u>procedure</u> being tested and not the individual bolt.

People often look at a chart that shows a torque value for a bolt. Those evil charts are common. People forget that there is a rule that one cannot go by one of those charts because the spread in actual pretension – the actual effect – is HUGE and unpredictable. The lower value of the spread would probably not even be close to what is necessary for an average bolt and the upper end of the spread may break an average bolt. So could one just take the middle value and assume it must be correct? No. That would be a code violation and is never allowed. One can **NEVER** skip Pre-Installation Verification because of a bogus chart or any other reason because Pre-Installation Verification is **MANDATORY** and there is **NOTHING** prequalified in bolting.

When doing the Pre-Installation Verification, another problem with this is that not only are you only qualifying the procedure one bolt at a time but also qualifying only that torque wrench or gun at that current setting with that current configuration and that current power source. That is not going to be the condition in the joints then or later the same day. It will be a different setting than what one will set on a truly calibrated manual inspection wrench. The joints will appear to be required to pass a different torque value during later testing, but it is really the same. One is just more reliable than the other.

Portable power sources are not constant and can be affected by the number of other tools they are operating, electrical or pneumatic (especially when someone parks a truck on the air hose). The power situation is impossible to predict. One situation at one brief moment at a load tester before work begins does not give a reliable indication of the settings required to achieve the minimum specified pretension at the joint later.

Things change every moment, and with a method which ignores the actual effect things can easily go bad while the contractor remains oblivious because the dial on the gun looks close to the setting that they may have used half a year ago or longer. What looks like the same setting never is, and so constant requalification or constant verification of the <u>effect</u> is necessary through methods such as periodically observing rotation or through torque testing with an actual calibrated and certified wrench. One reason that people think that they are being required to use more force during turn-of-nut pretensioning is because they get it closer to what it is supposed to be with that more reliable method. There are those who think that turn of nut gets is much tighter than required, but that is just not a reality. That is why such a reliability figure greater than 1.13 would have to be proven to be used, and there is little reason to think that a number like that is achieved in reality for torque-oriented methods with any consistency. It's not so much to be thought that turn-of-nut is tighter but that torque-oriented methods are so consistently inadequate by comparison.

One must use a few bolts to qualify the method and use an average value. But, as noted above, this method is grossly unreliable, especially when the contractor only attempts to do the minimum, which won't even be close. Reverification must occur frequently and it is often specified by the project engineer that the torque gun or wrench <u>must</u> be checked several times a day, even if no changes or adjustments have been made to the settings or configurations. The RCSC Spec also says requalification is required any time a change is made since any little thing can make it not pass testing, since the average (and by average I mean all) contractor is trying to do the minimum with zero margin or worse.

An easy way to verify it frequently is by observing rotation, assuming that one had the forethought to measure rotation at the time of establishing a torque value. This helps the reliability by observing the *effect* of the torque. And, observing the effect of the torque helps to link it closer to the Pre-Installation Verification testing where torque had no meaning. Torque has no meaning to a load tester because they only measure pressure exerted by a bolt on the plates. A load tester only knows what the bolt is doing and does not know anything about the wrench.

The Pre-Installation Verification is all about the effect of the procedure. Observing rotation will show two things. It will show when there is a problem with the torque gun or wrench, which can be frequent and occur throughout the day. It will also show how unreliable torque oriented methods are since the dial will have to keep being changed throughout the day even if no other changes were made. That can also be shown on the load tester. There may be a 20% spread in values from one day to the next for even the best gun, as it rattles around in the back of a truck.

It is also because of the Pre-Installation Verification that the "calibrated" wrench method, although in theory the same as snapping off TC bolts since it follows the same principles, may in reality be slightly less reliable than snapping TC bolts. This is because the contractors only want to do just enough to pass on the load test and no more. F1852 bolts, at least when treated right and all rules are followed, will snap off when they snap off, and no sooner than that, and the bolts won't be thinking "hey, this is good enough right here so I think I will stop and this must be tighter than this bolt will ever need to be." Those TC bolts, assuming that they passed the Pre-Installation Verification at all, will most likely have an average value even greater than the minimum required, although probably not by much. There is always something unpredictable about methods which focus on torque because one is always ignoring the effect.

If a contractor simply stops at the minimum only, then it is even more likely to fail any verification or inspection testing in the joint later, since there will always be the 3-8% relaxation under even good conditions. That has the effect of erasing much of the 1.13 reliability rating which represents only what is possible and not what is actual. Performing a test on a load tester has several defects which must be remembered to give it any value: 1) it is to be considered a perfect condition with controlled circumstances which can never be assumed to exist in a real joint, 2) it tests only one bolt at a time and does not test a joint of those bolts, 3) it does not take into consideration bolt relaxation over time. What may appear as meeting the "minimum" required by code may not actually do so in reality, and no study by scientists working in laboratories under controlled situations without real contractors doing what they do on a real project and without real inspectors ignoring what is happening will change that fact in the real world. They can only determine what may be theoretically possible to achieve. Torque can be very deceptive, especially when one is purposely ignoring the *effect* required by code, which is the only thing that matters.

The "calibrated" wrench method is more labor and time intensive than snapping F1852 TC bolts. So why use it? The bolts (hex bolts in this case) are much cheaper than TC bolts, and sometimes there are situations when only a wrench can be used on a bolt because of space limitations. There are also fewer rules than with F1852 bolts, although not by much. And, if done right, this method ("calibrated" wrench) can be reliable enough. It appears a little faster than turn-of-nut focused pretensioning, although that can be deceptive as noted above. Sometimes this method is specified for a project and the contractor forgets to submit a faster method. Price may be a factor, especially with some discount prefab companies which may supply cheaper hex bolts, but in the end are simply passing on the cost to the erectors who must then charge more for the extra labor – assuming that they noticed the situation in time to factor it into their bid – or take huge losses from time compared to fast TC bolts. Sometimes what may seem like a cheaper shop may involve hidden costs.

As with any other method involving A325 bolts, this method, compared to F1852 bolts which are what they are and can't be helped, one can still find a torque value which will produce the required tension during Pre-Installation Verification. One may observe a greater rotation for those 1" bolts than may have been expected, and they are more likely to cross the yield point than other A325 bolts (that doesn't apply to A490 bolts which only come in one strength) but one should remember that that is not a bad thing, it is not a good thing, it is just a thing and need not be worried about unless the bolts are clearly defective. 1" A325 bolts are softer than others within their required hardness range, but one need not compare them to other diameters which may be harder than required. The important thing is that they meet the requirements.

During this testing one would also establish an initial inspection torque for checking joints. This would be an initial value because the required torque can change over time, and if the project drags on, then one may need to establish a new torque under the new conditions.

TURN-OF-NUT

There are many projects where turn-of-nut measuring is used. The contractor will likely pull out a stupid chart showing that they only need to turn a bolt a specific amount and never any more than that no matter what was demonstrated on the load tester. They think that it applies to every bolt of every diameter with every length and every length of threads within the grip. I don't know why they put that chart in the RCSC Specification since it was never meant to be accurate and is only an example. The chart also does not distinguish between A325 and A490 bolts, which have different tension requirements. They, including inspector, forget, that it is not the rotation (elongation) that is the point. The pressure on the gauge is the point, or close enough to the point, during Pre-Installation Verification, since it represents the theoretically possible pressure on the joint created by the bolt. If a certain amount of rotation does not create that pressure then it is **WRONG** *no matter what the chart they say they looked at says.*

When testing for rotation one begins by having the contractor snug the bolt in the load tester to what the contractor will pretend is the same degree of snuggification which will be (or already has been) used in the joints. That creates the starting point for rotation, and is assumed to be the same which they intend to use. Here again it is lost on them that it is not the bolt being tested but their method of pretensioning, which includes snugging. It may be safe to have less snugness now, since then it will be a closer match to reality. They often try to put on a good show during this Pre-Installation Verification which will be difficult to maintain up in the air. They will snug the bolt in the load tester with *full effort* which will result in less required rotation than if they used the same effort they intend to actually use at the joint. The joints done to that lesser rotation are less likely to pass later testing since they are less likely to have been snugged to the degree used on the load tester. So, during the Pre-Installation Verification one should guess low on snugging, which will slightly increase the rotation required to produce the required pressure and will more closely match what they actually do. Also, snugged bolts relax and loosen as snugging progresses resulting in some desnugged bolts and variations in snugness. All bolts in the joint need to retain the degree of snugness qualified on the load tester for the next step to have any meaning. Only when what they do in the joint is the same as what they do when qualifying their procedure will it be likely to pass. This Pre-Installation Verification is not a game and not just a formality left over from some forgotten era and there is no place for pretending. Not following the procedure that they qualified is a code violation.

Something to keep in mind when qualifying the rotation procedure for each and every length, grade and lot to be used on that job is the joint the bolt will go in. Different lengths of bolts will be present for different thicknesses of joints. The only length that matters is the length between the plates. Also to be considered with that is the length of threaded and nonthreaded sections between the plates, which will affect the behavior of the bolts in the joint. The testing on the load tester should take such things into consideration through shimming or other method since the plates usually only come in one thickness, or two thicknesses if they are short-bolt plates. Getting the right bolt in the right joint can influence the success or failure of the joint by causing deviations from the assumed behavior tested in a load tester.

It is not a good idea to mix bolts of different lengths of threaded sections in the grip, such as a fully threaded bolt, in the same joint since the pretension may be different for each when using a rotation oriented method. Different bolts may have different proportions of threaded and unthreaded sections, from fully threaded to much less. This proportion affects the degree of rotation required for a given degree of tension. This is a noticeable factor even though a joint has the same grip all the way across (if you're lucky). A longer threaded length can give the impression of a little longer grip length or a softer bolt, since more of the strain is concentrated in the reduced section.

The stress/strain curve is exactly that – stress per unit area relating to a degree of change in length. The threaded and unthreaded portions of the bolt have different diameters. The threaded section only counts as three-quarters of the diameter of the nonthreaded section and therefore a given amount of stress (pretension) will produce a proportionately greater strain in that section than in the unthreaded section. It will strain an amount as though it was under a greater stress because the reduced area increases the stress density. The larger unthreaded portion has a lower stress density because the same stress is distributed over a larger area. This results in strain being concentrated more in the threaded section which has lower resistance to strain. The threaded section may even approach closer to the yield point in a way, especially since the stress is distributed unevenly on the threads engaged by the nut. Remember, the yield point is stated as pounds per square inch.

Therefore, different bolts with different lengths of threaded and unthreaded sections within the same grip will behave differently, and it can be seen on a load tester (since the load tester plates will have an identical grip for a particular diameter unless one shims it). Longer threaded lengths included within the grip will strain further with a lower pretension and may require more rotation (strain) to produce the necessary pressure compared to a bolt with a shorter threaded section in the grip. Yes, both the threaded and unthreaded sections have the same modulus of elasticity and have the same *nominal* diameter and follow theoretically the same stress/strain curve, but when the stress is divided by *effective* area of the sections it is like they are two different bolts in one. Changing the proportion within the same joint may have adverse effects because of different strain behaviors and different levels of pretension for a given degree of rotation, and one can prove it in a load tester. No chart, which people try to hide behind when they try to refuse to fix joints which failed testing, takes such things into account. In some situations a 1" bolt may behave more like a 3/4" bolt, and a ¾" bolt may behave more like a 1" bolt.

Because torque is related to surface pressure exerted on the face of the threads it ignores elongation. That pressure on the threads produces friction proportional to the pressure and requires greater torque as the pressure and elongation increase as one progresses with tightening. Because torque ignores elongation, it doesn't care about threaded and unthreaded lengths and focuses on the final pressure on the threads only (not the pressure of the bolts on the plates), after the bolt has stretched enough to produce such pressure on the threads. It doesn't care about how the pretension is distributed along the length of the bolt, but also doesn't care about what is going on *inside* the bolt.

Does that mean that torque oriented methods are more reliable than turn-of-nut? No. One can see that from all that has been shown in this work. Turn of nut, as an effect oriented method, focuses on the final result and one turns the nut until that result has been achieved. Turn-of-nut also uses torque like all methods do, but it doesn't stop there like the "calibrated" wrench method. There may be different factors which affect rotation and elongation, but that does not mean that it is unreliable. The geometric properties machined into the individual bolt are constant for that bolt, although these properties may vary by 5% or more between bolts through machining tolerances, and vary dramatically between different manufacturers. Charts and people who do not take all things noted above into account are unreliable. If it is done right, both torque and rotation should be equivalent no matter what you focus on.

Another thing not taken into account on those worthless rotation charts is the ratio of thread pitch to bolt diameter, nominal or effective. Different diameters have different thread pitches. As the size of the bolt increases (and the required tension increases proportionally to the size) the number of threads per inch drops, which means that one rotation produces a greater elongation. When one considers the number of threads per *diameter of length*, (which is what those charts supposedly focus on), which relates the thread pitch to the pretension level, the opposite of what one might expect from a greater elongation occurs.

It is remembered that the different bolts have the same modulus of elasticity and the same stress/strain curve. It is remembered that they are nominally the same grade within limits (when speaking of one ASTM specification of course), and even when comparing bolts from different specifications, for instance A325 and A490, hardness can be neglected in the elastic range. It is remembered that the full pretension is 70% of the tensile strength per unit area. A given diameter is therefore related to a particular pretension. When considering thread pitch one can compare the thread pitch to diameter and see 6.5 threads per diameter for a 1/2" bolt and 8 threads per diameter for a 1" bolt. Even though there is a greater distance between the individual threads as diameter increases, the number of threads per diameter is also increasing, not decreasing. Because diameter relates to a particular load one can consider the thread pitch to diameter ratio as equivalent to a thread pitch to pretension ratio. One can see by this that as the bolt gets larger and the required stress increases, it will appear as though the larger diameters are softer and require a greater rotation than smaller diameters, and above 1" for A325 bolts they are softer as well. This is considering *diameters of length*, which is also what the RCSC Specification considers in the chart. As diameter increases, so does the pretension, and so does the elongation required to produce that tension. But when considering thread pitch, that increase in rotation from one diameter to the next is not linear. Comparing 3/4" bolts to 1" bolts, to make it linear would require that 1" bolt to have 7.5 threads per inch, or a thread pitch of .133" rather than .125". But, they are not made that way, yet.

That is considering *diameter lengths*. Is that fair, since longer bolts have more metal to stretch? No, but it is *diameters of length* that the chart goes by and a diameter of length is longer for a larger diameter. What about the case of two diameters being used with an equal grip length within their respective joints (different joints of course)?

We see a different situation if we consider equal grip lengths, such as when two different diameters are used in similar joints with the same thicknesses, ignoring rotation. Here, since modulus of elasticity is the same for all bolts no matter what grade or they happen to be, and since diameter is directly related to pretension, we see that a 3/4" bolt and a 1" bolt would both require the same elongation to produce the 28 kips and 51 kips respectively, if all other things are identical, since the stress density is identical for all diameters of a particular grade. That is in theory only and ignoring other factors that make such a coincidence unusual in reality. The rotation required to produce that elongation then depends on the thread pitch. A full turn of a 3/4" bolt relates to a lower elongation (0.10" in theory only) than the slightly larger thread of a 1" bolt (0.125" in theory only). A smaller diameter bolt has more threads within a given joint thickness. Each thread moves the nut a smaller distance for a smaller pitch. Since the total required elongation in this case is the same, we can see what seems more intuitive -- that one bolt will reach that point sooner than the other in terms of angle of rotation. This is not considered in the useless charts which consider only *diameters of length*. This is an oversimplified description which ignores the uneven distribution of stress densities over the *length* of the bolt which influences the total strain of a bolt for a given stress level, and ignoring many other things. But, turning a nut a full thread or any fraction thereof, for a specific bolt, will always produce a given strain depending upon the thread pitch, which total strain is equal to that fraction of a thread, even if it is not shared equally by all sections of the bolt.

Of course, in the real world and with usual design practices, the joint thickness and bolt diameter are at least partially related, with bigger things using bigger bolts, but that is often not the case. One can use a larger number of smaller bolts in a joint in many cases, since the design of a bolted joint is based on the load to be resisted (such as shear) or transferred (such as tension or compression) and is not so much related to the thickness of the plates, except to the extent that the load determines the plate thickness at each joint and the practice of using a standard all-purpose design thickness is not used.

It may not be necessary for either the inspector or an ironworker to do the math only slightly touched upon above. It doesn't change the reality one will see on a load tester. What does matter in this discussion is that there is no chart one can use to eliminate Pre-Installation Verification testing. To use a chart as anything other than a cute example of what may be possible for educational purposes only would be a violation of the intentions and the letter of the RCSC Specification, even though someone had a lapse of judgment and included that chart in the RCSC Specification, because that Specification <u>mandates</u> testing and that the contractor <u>must</u> follow the procedure qualified by that testing, no matter what any chart says. They included that chart as an example only, not thinking that contractors (and even some inspectors) would try to use it to ignore or to even protest against the <u>mandatory</u> testing requirements or even the actual results of such Pre-Installation Verification testing. One *should* be able to expect anyone and everyone to understand that there **NOTHING IS PREQUALIFIED IN BOLTING!**

Another indication of the inapplicability of those charts and which makes a difference with rotation is the fact that there will be a different angle of rotation for A325 and A490 bolts. That is very clear from looking at the stress/strain curves, which have identical slopes in the elastic range but very different required pretension levels.

The biggest difference between A325 and A490 bolts and their derivatives is hardness and strength, especially with diameters over 1", where A325 bolts have a lower required strength but A490 bolts have the same required strength for all diameters. That hardness itself does not make any difference in the elastic range since both types, and all types of steel in theory, have a roughly identical modulus of elasticity. So, within the elastic range, an identical elongation (rotation in this case) produces an identical stress (tension). But, that is only in the *elastic* range of both bolt types.

Where that elastic range ends and the nonlinear plastic range begins (yield point) is very different. One bolt (A325) will reach that point long before the other (A490). And especially for larger diameters, an A325 would have not only yielded but may have already broken when an A490 begins to yield. That difference is clearly visible when comparing the stress/strain curves. One extends much higher than the other.

A490 bolts require a greater tension than an A325. Since the elastic ranges of both overlap and the modulus of elasticity is the same, then the elongation which produces, for example, 51 kips in a 1" A325, would in theory also produce 51 kips in a 1" A490. But, an A490 can't stop there. A 1" A490 must continue rotating until it reaches 64 kips, which is the required *minimum* pretension, which is approximately 70% of the tensile strength and close to the yield point.

Therefore, one can see, even by looking at the stress/strain curve, that they will obviously have different angles of rotation. The chart method fails again!

F1852/ F2280 METHOD

As with all other bolts, so too with F1852 and F2280 snap-off bolts. When testing, one must begin at snugtight. One might not think that snugness would affect the final result, since this is a friction based method, but it does, especially for bolt groups where these bolts relax as bad as any other method not done right. A significant initial tension will result in a slightly higher final pressure on a joint after all the bolts have been snapped, even though one does not see the effect as clearly on a single bolt in a load tester. Better snugging means lower loss of tension in bolts previously snapped. That should be a warning to contractors that it is necessary to follow procedure and snugging is part of that procedure. All too often (actually always) people think that snap-off bolts are somehow guaranteed to always be right and no one has to think about anything or even try to do it right.

The Pre-Installation Verification testing of these bolts also suffers from the flaw of only testing one bolt at a time and it is not possible to test an entire joint on a load tester. When one bolt is snapped in a joint, it is just a snapped bolt. It is the total joint which matters, not just each bolt as an individual. The entire joint is dependent upon each bolt not only being pretensioned (which of course must occur one bolt at a time) but maintaining that tension afterward. That means that with this method, as with every method, every bolt must be snugged first before any bolts are snapped.

As with all methods, Pre-Installation Verification of this method on a load tester suffers from that fact that it is being tested under perfect conditions. Also, as with testing any method, it is not possible to test *assemblies* or *joints* without more advanced and expensive methods than are likely to be used, but only one bolt which is all people can fit in a usual load tester. That can give a false impression that proper fit-up and snuggification of the joint are less important. But, plates must be in alignment and in *Firm Contact* before proceeding. The Pre-Installation Verification on a load tester, like every method, really only qualifies a procedure for use on a load tester, and not necessarily a real joint. It only qualifies that procedure for use on a joint which meets the same perfect conditions as they exist on a load tester. Any movement of the plates in the joint, even if it is so small you don't see it, removes the pretension from the bolts if they had already been snapped. A movement as little as 1/32" can be the difference between tight and loose, since the difference between snugtight and pretensioned can be as little as 1/3 of the thread pitch, which is small to begin with. Even a snap-off bolt will have the same rotational requirements as any other method if it is done right. If pressure shifts in the joint as tightening progresses, then there will be bolts which are no longer properly pretensioned. And, once they are snapped they can't be resnapped. Therefore, it is very critical, as with every method, to make sure that every joint is fit and aligned properly and every bolt is snugged properly before proceeding.

This method is torque oriented and suffers from the usual torque oriented spread in actual values which one hopes will average to something passable. Everything which affects torque affects these bolts. If these bolts are not taken care of and stored properly they will not pass. One can't just turn the TC gun until it is right. One can't adjust the torque setting on a TC gun. The bolt will snap at a particular torque value and no more than that, whether it is right or wrong. The bolt does not snap at a particular pressure. There is no adjustment which can be made to the bolt.

In that particular sense, there is less you can do to help one of those bolts. But, it is still perhaps more reliable than the "calibrated" wrench method *because* it snaps only when it snaps and no sooner than it snaps. The TC bolt doesn't care about effort and neither does the ironworker pulling the trigger since the torque is not felt. A TC bolt won't stop early and won't be trying to do only the minimum and no more, although that can quite often be the result because of the huge spread in values. The "calibrated" wrench method usually does even less and is less likely to *maintain* the pressure.

When sampling and testing, it must be tested as an *assembly*, with bolt, nut and washer(s) installed by the manufacturer. It is only the entire assembly which passes. One cannot take a nut from somewhere else just because someone dropped a nut. That would make it a completely different assembly which would have to be qualified. The manufacturer will only approve and guarantee the use of assemblies as they came out of the factory. That is the end of the manufacturer's liability. The success of the assembly depends as much on the nut as it does on the bolt, since the nut is also threaded and it is the nut which applies the force for both elongation and snapping.

ONE FINAL METHOD: DTI WASHERS

I will leave it as an exercise for the reader to determine why I don't need to discuss direct tension indicator washers here.

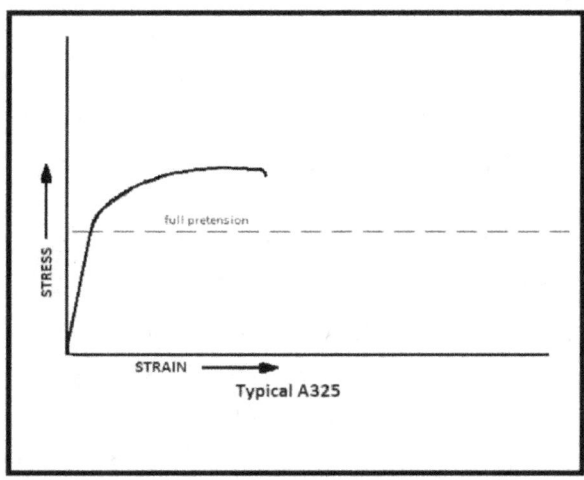

NO FURTHER VERIFICATION? WHEN?

"It's always right because I've been doing it that way for 30 years!" Even if that is in dog years, as I suspect, they should know how to do their jobs. This includes inspectors. Some method can never be considered right just because they, supposedly, got away with it for so long. At some point in their career, they need to stop and make sure that they are doing things right, no matter what someone told them once or what they have gotten away with.

Many times an inspector is called to the jobsite for bolt inspection and upon arrival finds that erection and pretensioning had begun some time ago, or may even be completed! The contractor or project manager then uses the excuse that *all* bolt inspection is "periodic" only, and also will try to claim that "periodic" means absolutely no inspection whatsoever and any stage. They ignore code requirements, and part of the problem is that so many inspectors also ignore those same requirements. If inspectors did their jobs, so would the contractors. Only then would they not be so confused when a **REAL** inspector arrives.

WHAT ABOUT PRIOR VERIFICATION? Show me the report for the mandatory Pre-Installation Verification which must occur before any erection begins and must occur at the same time that the cans of bolts are opened for the *first* time. Show me the report for the inspection of all surfaces showing that they meet the requirements for condition, including slip-critical joints where specified. Show me the report for the inspection of erection, fit-up and snugging. Prove to me that while no one was watching during the past few months, or even longer, that only bolts sampled and tested and passed during the mandatory Pre-Installation Verification were used and absolutely no others. Show me the report from an inspector which shows that that the right bolts with the correct threaded length in or out of the grip were used in all joints so the nut doesn't hit the thread runout, and the threads are excluded from the shear planes where required. Show me the report that states that any reaming was dealt with and not simply covered up. If any of those, and more than those, are missing, then it can't be passed by an inspector. Too bad! So much for "periodic" meaning absolutely no inspection or verification of any kind at any stage of the process, each stage of which requires verification before proceeding.

What are we looking for? We are looking for something which cannot be seen directly. We are looking for something which occurs *inside* the joint between the plates, completely concealed from view. It is not the bolts themselves that we are most concerned with but what those bolts are accomplishing. The bolts are only a tool which makes the success of the joint possible under only the right conditions. How can we know that which we can't see on the outside of the joint after the fact? We can only estimate the possibility of the existence of the required clamping force by <u>observation at EACH stage of the process</u>. If one thing is missed, then it is not right, no matter what it looks like on the outside (which means absolutely nothing) and people may die. The only way that bolting can be done right with success is if every step is done exactly right and the procedure determined during the mandatory Pre-Installation Verification is performed exactly and verified.

There is nothing "prequalified" in bolting and it is unlikely that will change any time soon. There is no perfectly reliable method, and the people performing it and inspecting it are even less reliable. There lies the problem. The bolts are what they are and their success or failure depends completely on how they are treated and used (that assumes that they passed the identification, sampling and mandatory Pre-Installation Verification and anything with manufacturing defects eliminated). Any good, perfectly manufactured bolt can be used wrong and fail, sometimes even catastrophically.

Was it then a bad bolt, if it failed? Were the expectations of the contractor realistic (not to mention those of the engineer)? If it was a bad bolt, then why was it used? Any bad bolts would have been noticed during the mandatory Pre-Installation Verification which really is MANDATORY since it is impossible to do that after the bolts have already been used. Were the bolts taken care of and stored properly and protected from damage and contamination? It is often the contractor and inspector who fail, and a collapsed building or bridge is the result.

Who is to blame? Who is liable when the bolts break because the contractor believes that loose bolts are better? Should one blame the manufacturer because the contractor used their product improperly and refused to follow the specifications? *PEOPLE FORGET THAT THE CODES AND SPECIFICATIONS WERE WRITTEN IN BLOOD*, and it is stupid and criminal for people to assume that they know better than the scientists who studied those failures and are trying to prevent them from happening again. Lack of knowledge kills.

How do you know if it is right, and by right I mean that the code required pressure exists between the plates? One obviously must start doing things right before erection even begins. The Mandatory Pre-Installation Verification and sampling have been discussed enough for anyone to understand that it is Mandatory and nothing can be accepted later if it is missed. Nothing is prequalified in bolting and no method can ever be assumed to produce the required pressure between the plates unless the specified conditions have been met and verified.

Nothing is prequalified in bolting. No method is capable by itself of producing the clamping force needed to hold a joint together and transmit the factored loads unless *all* things are done right. No method excuses negligence. No step can be skipped. All methods are highly dependent on all conditions being met, and verified that they were met, or rotation, torque, snapping or crushed DTI's don't mean anything. They may show that effort was involved, but that has no meaning *ever* unless *proven* to effectively produce and retain the internal forces which can't be seen afterward or even tested for. (There is a method of ultrasonic elongation measurement which can show that tension exists in a *bolt* – assuming that testing was done previously on those bolts to calibrate that method during the Pre-Installation Verification – but even under the rarest of cases when someone goes to that absurd expense, that still does not show the pressure *between* the plates, only the potential pressure *outside* the joint from the tension inside the bolt.)

Nothing is prequalified here. Even the Pre-Installation Verification can only determine what is *possible* under perfect conditions. Do the joints meet those conditions? Have the joints been inspected to verify that they meet the required conditions? Were the joints clean? Was there any reaming necessary? If there was reaming, are plate washers or even welded plate washers required? Is the joint already fit with the bolts already in place? If so, then how can one see that the joints were clean, that they meet slip critical requirements if necessary, or that welded plate washers are not required? Obviously one must be observing all that and verifying those things in progress.

If the holes or surfaces of the plates are not in compliance, then the joint will not perform safely. Were the holes gouged with a torch without approval? Were they reamed afterward to eliminate gouges, if it was approved? How is it possible to know unless it is inspected <u>before</u> the bolt covers it and hides it from view? Are the holes slotted now, when they were supposed to be standard holes before reaming? Then the joint, if it is allowed by the engineer at all, <u>must</u> now be treated and inspected as slip critical or plate washers must be welded to make them standard holes again. (Note also that the increased joint thickness at the plate washer will have an influence on required rotation and influence how the bolt shares a load with other bolts in the group, because there will be more bolt length included in the grip to stretch and will give the same elongation as the other bolts in the joint but at a lower level of pretension.) It is not the tension in the bolt that is the point, but the effect of that tension on the joint and the ability of that joint to remain functional under load. Anything which would allow slippage or make the relaxation of the joint worse under load makes that joint noncompliant no matter how tight the *bolt* was at one time.

Was the actual installation of the bolts observed to see that no bolts were damaged? Occasionally an ironworker may even try to hammer a bolt through the joint, thereby damaging the threads. If a torque oriented method is used, then it is impossible for a bolt with either damaged or unclean threads to reach the required pretension. Was there any bolt "trapping" taking place from sloppy erection practices and shifts in the joint exerting severe pressure on the bolt? If so, that bolt is doing absolutely nothing no matter how tight the nut is on the outside.

Were only bolts identified, sampled and tested and passed used in the joints? How do you know if you don't observe the bolts being removed from the cans and used in the joints? Bolts are not individually marked with lot numbers embossed on the heads. One can only witness the removal of the bolts from the identified and properly protected cans, observing the condition of the bolts.

Were the correct bolt lengths used at each joint? Are the threads (only roughly three-quarters of the nominal diameter plus or minus) allowed in the shear planes of the joint (marked "N" on the plans)? Is it possible to see inside the joint after it is fit, to know that the threads are excluded where required (marked "X" on the plans)? How does one know fully threaded bolts were not mixed with others of the same length in the joints unless one saw the bolts installed?

Bolts come in different lengths with different proportions of threaded to unthreaded lengths, (the threaded length being the same generally), and for some diameters of A325 they can come fully threaded. Many different bolts can be used on a project. Which one was used where? Did they install the bolts with the proper position of threads to avoid hitting the nut on the thread run-out, which would make any pretensioning method worthless and unreliable? (It is just as bad if they use bolts with damaged threads.) When a nut hits the thread run-out, a torque wrench will still click and an F1852 bolt will still snap off, but it won't have the required tension and the joint will not have the required pressure between the plates, which is the only thing that counts.

We are looking for something completely invisible until FEMA investigators come in to determine why the building collapsed when the contractor insisted that the joints are stronger than they will ever need to be (which they do every time, even when they won't do anything they are required to do, and including the many times when you can spin a bolt with your fingers in a supposedly already pretensioned joint). The only thing we can do is observe <u>every</u> step of the process to determine the potential for the bolts to create the necessary pressure within the joint. It is only possible for that invisible pressure to exist if <u>all</u> things are done correctly at every step and <u>all</u> things are verified while that is still possible. We know what must be done to make a joint work and it was demonstrated in the Mandatory Pre-Installation Verification, and we know that not doing those things exactly will not give the results shown in the Mandatory Pre-Installation Verification. Also, how can one know later if F1852 bolt *assemblies* have not had components of different lots mixed, which disqualifies them and voids any liability of the manufacturer if there is a failure? These things can't be tested for later when the bolts and washers are covering the problem.

One is not required to track which joints a particular lot number is used. That implies something very critical which is normally ignored. It can be believed by contractors, inspectors and project managers that it means that inspection is not required until *after* installation and *never* before. That can be used by the less observant to try to excuse skipping the Mandatory Pre-Installation Verification, even though they have no excuse for not understanding that testing is required before they install any bolts, as the name proves. The fact that one is not required to track each lot in use actually proves that the inspector is <u>required</u> to be on site and involved from the very beginning! It proves that the codes and specifications for steel construction actually do <u>require</u> and fully intend that the Pre-Installation Verification is Mandatory, to have the inspector present from the beginning, and that it can't be done after the bolts are in use!

Let's look at what it means. It means that the Pre-Installation Verification, which is Mandatory no matter what anyone wants to try to claim after they get caught, has already occurred, all bolts to be used on the project have been sampled, identified and tested, and that there are no nonconforming bolts left on the jobsite. Any bolt not in compliance for any reason, including both manufacturing problems and improperly stored and mistreated bolts, has been eliminated. Because all bolts which are now still at the jobsite have passed inspection, then as long as no new bolts are brought to the jobsite, which would require further testing and sampling of those, then every lot being used has been sampled and tested and is compliant. The inspector *knows*, and only under these circumstances can the inspector know, that all bolts in the joints are identified

It is obviously in the contractor's best interest to make sure that the Mandatory Pre-Installation Verification and all necessary inspections occur before using the bolts, because otherwise it is impossible for the inspector to know if all the bolts being used have been sampled. Is it possible for the contractor to *prove* that all bolts used before an inspector was called came only from the empty cans still left on the site and sitting by the pallet, or pallets, which is far too large and too many for one can of bolts (also obvious by the quantity of plastic wrap supposedly wrapped around only one can on that pallet)?

Is it possible to sample bolts from the joints, even if they have not been pretensioned yet? *Every* joint would have to be sampled, since it is impossible after the fact to know how many lots are involved and where each of those went. Every joint would have to be sampled to make sure that every lot is represented. The project can't proceed like that. It is a very costly mistake.

What if a problem with the bolts is discovered later, which would have been prevented if the contractor followed the rules? Every bolt similar to the bad ones would have to be removed from the structure and replaced, which would not be necessary if the contractor followed the rules and the inspector was there from the beginning as required. One can't look at a bolt and know what lot it came from, and the cans get thrown away when they are empty, or even combined. One can't know that a bolt came from a particular can unless one observed it come out of that can and that the only bolts in that can came from the supplier of that can. One can only remove all similar bolts to make sure any of the bad ones are removed and eliminated and none remain anywhere.

Do all the bolts in a joint match? This is a matter which is sometimes visible in the stickout. We want to assume that someone was also not stupid enough to combine different diameters in a joint (such as when they can't fit the correct size bolt through the holes because they don't line up – yes it happens). Are the bolts the correct grade specified for the joint? (We want to assume that they are not stupid enough to combine different grades in the same joint because they ran out of the correct grade.) The bolt assemblies, if we may call them that, are also only as strong as their weakest member. Are the nuts the correct grade and heavy hex? If not, the chance of failure is immensely greater. A nut which is too soft may yield long before the bolt does, and they will not share the load because of different behavior at the stress levels. One may see the different lengths of bolts in a joint as the ironworkers grab whatever is on hand at that moment and you can see it in the stickout past the plate and even in the shims used with that longer bolt, which should not occur if the correct length of bolt is used in a joint. Each bolt can behave slightly differently if there is a difference in the length of unthreaded sections in the bolts of a single joint within the grip. That is pretty much ignored in the specifications, but one should be able to assume that it is not necessary to say not to do that. It is not enough that an individual bolt theoretically demonstrated the ability to meet requirements in a load tester. In an actual joint the bolts must work as a team. One bolt could in effect take a proportionally greater load than others (we are talking about those situations where that was not created from eccentric bolt patterns or faulty design), and it won't be shared evenly by all the bolts. There is little said about such things in the specifications, but there can be a difference, and it would seem that one would want the joints to perform at their best and reduce liability, which can only happen if the bolts can perform predictably. It is also not possible to address in the specifications and codes every stupid thing people may try to do, since people are always coming up with new ways to do things wrong.

NO FURTHER VERIFICATION REQUIRED? What about prior verification?

What if pretensioning is not required? All that has been said applies, except pretensioning and what applies only to slip critical joints. The steel must fit together right. That should be obvious. The joints must still be clean. Holes still need to be right. Reaming still occurs and must be addressed, and oversized holes and slots which occur from erection practices become even more critical to the success of the joint than if it was a slip critical joint (and might now have to become one) because of the small amount of friction produced in a snugtight joint. The correct specified diameter of bolt must still be used. Joints must still be aligned right.

For even snugtight joints there are verifications to be made and inspection of the bolts and steel is required, except without determination of rotation or torque to bring the bolt to 70% of its tensile strength. But – they all try to say – there is "no further verification required." Does that mean that there is absolutely no inspection at any stage? One should notice that there are verifications to be made without regard to whether or not the joints will be pretensioned. One may point to a note which occurs *after* the required verifications in the section on snugtight joints which does say "no further verification required."

Let's look at that statement. The word "further" implies, as should be obvious from the fact that there are verifications listed as required, that inspection from the beginning up through the actual snugging has occurred. It means that the inspector has been involved in the project from the beginning. It means that the steel has been inspected. It means that the inspector observed that the bolts have been snugged and that the plates won't move or the bolts loosened as snugging progresses, which is always required whether they get pretensioned or not. It means that the inspector saw them using <u>full effort</u> as required for snugging just like for the pretensioned joints. It means that the inspector saw that the plates are in firm contact and there is no potential for movement which could loosen the joint. It means that the inspector observed the bolts snugged in a manner which would minimize any relaxation of previously snugged bolts. It means that the inspector observed the correct lengths of bolts used in the joints to avoid hitting the thread runout, which would cause a bolt to remain loose and would not meet even snugtight requirements and would allow the joint to be damaged in service. It means that if reaming took place and washers were required that the inspector saw them installed where necessary before any bolts covered the area. It means that the inspector saw that there was no thermal cutting of holes without approval. All things required for all joints up to this point have been verified. <u>Then, and only then, verification of pretension is not required in joints that are not designed to be pretensioned</u>. Then, and only then, one may stop, having observed all requirements for those joints and there is nothing further to verify.

Where does it say that there is no inspection ever? Where in this does the word "periodic" apply in the sense it is normally interpreted? It is only at the very end of the process that snugtight joints differ from pretensioned joints. There is no difference otherwise. Because of all the verification requirements and because pretensioning is not required for snugtight joints, inspection of snugtight joints is effectively continuous on common projects, since the only things really allowed to be inspected in a noncontinuous manner without causing delays in the work are the final stages of only certain pretensioning methods, which stage is only reached *after* the effectively continuous inspection of what we have been discussing here, which is required for all joints. I say "effectively continuous" since that is what it ends up looking like no matter what people call it.

People think that snugtight joints don't matter. The difference between tight and loose is even finer for snugtight joints, and the relaxation takes away a much larger percentage of tension than for pretensioned joints. Snugtight joints are far less reliable than pretensioned joints. This is because the definition of snugtight is quite arbitrary, and at least with pretensioned joint, they come back and tighten them again after snugging. There is no arbitration torque one can establish without the Skidmore testing performed with pretensioned methods and observation is the only means of verification for snugtight joints. Pretensioned joints may still be tested for compliance later. People often don't even try when it comes to snugtight joints and the bolts can often be spun with one's fingers when it was supposedly tightened to the *full effort* of the ironworker, which can actually be up to a couple hundred foot-pounds depending on the individual and the wrench used. If a joint is not properly snugged, the building and items attached to the building are in danger of damage from the shifting of the joints. The reasons that some joints are allowed to be snugtight are 1) no seismic or dynamic loads or vibration, 2) shear only with no tension or only very small static tension on a less important joint. The level of pretension in a bolt has only a minor effect on the shear capacity of a bolt, but that assumes that it is not loose. A loose joint can rotate when separation occurs and the joint can shift much further than the hole clearances. Bolts can become even looser over time and may even be out of contact with the surfaces of the plates. So, even for snugtight joints, it is tension which protects the bolts and joints, even if that tension is much lower.

The preceding discussions should make clear the irrelevance of thinking in terms of "periodic" or "continuous". The effect of the requirements makes these inspections have a very continuous character, or it becomes impossible later. It hardly even needs to be specified as "continuous". If there is any inspection done correctly then one can hardly talk about "periodic". Things either get inspected at the necessary stages or they don't. Erection is a continuous process and inspection must match it. There is only one stage where such a distinction is possible, and on most jobs it still requires the inspector to be present continuously because of the volume of work and the fact that one area must be pretensioned and passed, such as a lower floor, to move on, and not everything happens at once. The one time stage where there is *sometimes* a distinction possible, in theory, between "periodic" and "continuous" is at the end of only certain pretensioning methods and even then under only certain conditions.

If a pretensioning method can be allowed to be inspected "periodically" under certain specific circumstances, that does not mean that there was no inspection up to that point, or that the inspector was not required to be there every day to prevent the failure of the inspection process. Certain methods only become "periodic" at the very end of the process, and we will take a look at those. As shown above, the inspector must have already been involved in the project effectively continuously up to the point of pretensioning, and all things verified which are required to be inspected at every stage and which can affect the success of the joint in use. It should be clear that if the contractor feels like doing all that is required, then, because proper snuggification is the foundation of all joints, the pretensioned joints obviously must require prior verification of all things before any bolts are pretensioned. Obviously, the "no further verification" statement has no meaning in pretensioned joints and only refers to joints where no pretensioning is required. If the joints will be pretensioned then the inspection must continue. Sometimes rebar can be inspected "periodically", but after the concrete has been poured it's too late. Every stage of bolting is like that.

In those cases where pretensioning is required <u>after all which has been stated above has been verified and inspected</u> and all joints have been snugged properly, then it is time for inspection of the pretensioning.

Just because certain methods under certain conditions and only on certain projects inspection may possibly be "periodic" <u>never</u> means that *all* inspection is *always* "periodic". It also <u>never</u> means that the inspection is *only* to occur after pretensioning when it is then impossible to verify all the many things which must be inspected. There was a procedure which must have been submitted and approved by the project engineer. This approved procedure must have been tested during the Mandatory Pre-Installation Verification to determine that it has the *potential* of meeting the requirements under the specific conditions of the test and that the contractor is able to perform it correctly. It is only capable of meeting the requirements if it is followed exactly. It is expected, <u>and required</u>, that the inspector verify that the contractor is following that procedure without deviation, since any deviation would be a code violation, unless more testing is performed to qualify that deviation. How is that to be done? Obviously direct observation is required, especially when so much is impossible to verify later until the building collapses, and especially also when the pretensioning methods used *require continuous inspection.*

Let's take a look at some of the methods and how the word "periodic" is interpreted by the RCSC Specification.

Turn-Of-Nut: periodic or continuous inspection?

How are they monitoring and controlling the rotation of the nuts? Are they even turning them the amount determined to be necessary during the Mandatory Pre-Installation Verification? Who is paying attention to the rotation? What evidence is there of the correct angle of rotation?

Turn-of-nut pretensioning, contrary to what most people try to say after the fact when they get caught, always requires continuous inspection per the RCSC Specification, unless one condition is met to make it "periodic" at one last stage. That condition is matchmarking, where the rotation becomes visible and remains visible after the fact. Otherwise rotation must be directly observed by the inspector.

Let's say that they choose to use matchmarking, which is quite simple and fairly reliable and provides the ironworker with a visible indication of compliance with the procedure. Does that mean that there is *no* inspection until the end? Is that what "periodic" means? No. No such thing is ever intended or could ever be inferred from the word "periodic".

When matchmarking is used, the inspector <u>must</u> witness the marking of the bolts <u>after</u> witnessing the snugging and <u>before</u> pretensioning, or those marks have no meaning, and it is the entire joint which must be snugged properly first and not just a single bolt. Anyone can make any mark when no one is around, which could make a loose bolt appear pretensioned. It is only after the inspector witnesses the marking of a properly snugged joint to see that they have been properly marked per the procedure that the inspector does not need to witness the rotation of each bolt. It also is obviously necessary that proper fit-up and snugging was performed and verified before marking. "Periodic" clearly intends the inspector to be involved throughout the process, and never only at the end after the work is completed. It can be seen that "periodic" has a very limited meaning with this method, and if the contractor does not want to follow every requirement then it must be inspected continuously at every joint through every stage.

On most projects "periodic" has no meaning. Every bolt would have to be marked and the contractors won't mark the entire building at once and then pretension the entire building at once after the inspector leaves. The inspector will likely have to be on site on a continuing basis to verify the marking so that they can be turned with only "periodic" inspection at a particular individual joint. It is hardly necessary to specify continuous inspection on a project – it is either inspected right or it isn't.

Direct Tension Indicator Washers: periodic or continuous inspection?

The idea behind the use of these washers is that it takes a certain amount of force to crush the bumps to a certain gap or to squirt the paint. It may even be considered more reliable than turn-of-nut, when done exactly right and all things have been verified before pretensioning begins. Does the potential reliability under the right circumstances mean that there is to be no inspection until after it is completed? The RCSC Specification never intended such an interpretation.

After all the bolts have been fit and snugged with the DTI washers in place, the inspector is required to check them and make sure that the washers meet the requirements specified before pretensioning. That means that the inspector is required to be present for the installation and snugging of the washers. That means that the inspector is required to be involved <u>before</u> any pretensioning begins. The washers must be installed correctly and verified first. If the washer is positioned incorrectly, it must be removed and placed correctly before proceeding. There are requirements for how the washers are to be installed, and if those requirements are not met, then there is absolutely no meaning if the gap measures correctly afterward. Only after inspection of the installation can the contractor proceed with tightening.

Observation and inspection of the installation is critical. Bad washers may get used. Cheating is very common, especially when the contractors won't let anyone know that they are proceeding. Bumps can be rubbed off through intent or through improper installation. Squirted paint may mean nothing more than that the ironworker also carries a hammer, not just a spud wrench. The conditions of the washers – if supplied by the contractor – must be verified to be compliant before use. The effect of all this is to make DTI use no less continuous than any other method.

Also, it must be remembered that a crushed DTI washer only shows that the pressure reached a certain level at one time and does not necessarily imply that it still exists, especially with the paint types. At best, it only shows that there was pressure applied *outside* the joint. It is just as possible to tighten the bolts with an improper sequence which can cause relaxation of previously tightened bolts (or make that relaxation worse). Cooperation with a knowledgeable inspector is the only way the joint will be safe.

Is this "periodic" only? To meet the actual inspection and verification requirements for this method the inspector will usually be there every day on most projects since buildings are built a section at a time, and "periodic" has no real meaning if it is done right, or if the contractor wants to proceed as efficiently as possible.

DTI washers help when it comes to dealing with the short bolts which exist on every project and which can't be tested in a load tester without the plates for short bolts.

Only after the inspection of the installation of the washer can the contractor then proceed to pretension the bolts without the direct observation of the inspector standing right there. It is only the final stage where "periodic" has any meaning for an individual bolt.

Torque Wrench – "Calibrated" Wrench method: periodic or continuous inspection?

This method is only to be inspected continuously, and one should not try to use the idea that the bolts can be torque tested later because the contractor always loses and the project is delayed since every bolt will end up having to be tested, usually repeatedly. It is far more efficient to have an inspector there than to do it twice (or more). This method is <u>not</u> allowed to be inspected "periodically" under the RCSC Specification. It is grossly unreliable anyway.

F1852 and F2280 TC Bolts: periodic or continuous inspection?

Even these bolts must follow all the same rules as all other methods, and also some which apply only to them. These use an unreliable torque method which is as highly variable as the "calibrated" wrench method. These bolts are very sensitive to conditions, as are all pretensioning methods.

It is obvious that the Pre-Installation Verification applies as much to these as to all other bolts and methods. It is a step which can't be skipped. Sampling must occur <u>before</u> any use, which is the only way to know that all lots used on the project are represented. If it is not done <u>before</u> installation, then one can't know later that the bolts left over are the only ones used. Clearly this is no different than with every other type of bolt and method.

Proper fit-up and snugging is critical. Proper care of the joints and holes and bolts are critical to the performance of the joint, and those are things one can't see later. Those bolts, which are falsely believed to take care of themselves, will relax excessively if <u>all</u> bolts in a joint have not been snugged properly first, and if they are not snapped off in the proper sequence. A snapped bolt **NEVER** implies that tension exists unless every step of the process is verified before snapping them. That means that the inspector must be present and involved from the beginning to verify that the proper procedures are followed. The only stage where it could possibly be allowed to be "periodic" – assuming total cooperation from the contractor at all stages which is rare – is at the very end when a bolt is snapped. For that to be "periodic", everything must have been verified at the joint by the inspector up to that point, since those things can't be verified later. After all things have been verified first, then the contractor can snap the bolts without observation. We are looking for something completely invisible. It is impossible to see pressure *between* the plates. It can *only* be *assumed* when all things have been verified at each stage, and on most projects what contractors believe to be "periodic" inspection will require the inspector to be present continuously to verify all things necessary at each joint, or the project will be unable to proceed.

Because we are looking for something completely invisible, and it is not possible to directly test the pressure between the plates produces by the appearance of things outside the plates, we can only make assumptions based upon known facts. The inspector is not so much witnessing the pressure between the plates, which is impossible but which is the whole point of bolting. The inspector can only directly verify that the procedure was followed which is known to be capable of producing that invisible pressure.

It can be seen that the only times that the word "periodic" has any meaning is only at the very last stage of only certain pretensioning methods and not other methods. Every stage leading up to that has verification requirements which must be met before proceeding. All those other steps must be verified before pretensioning, since those are things that one can't check afterward. For something to be literally periodic, it must be possible to verify compliance after the fact, and that does not apply to most of the process of high strength bolting. Even in those cases where one can inspect on a "periodic" basis, it is only the very last step of only certain methods that one can see later, such as matchmarking or snapped off splines, and observation of those two things never implies that the other requirements have been met. If something can't be checked later, like all steps before pretensioning, then the inspector must be there on site for it.

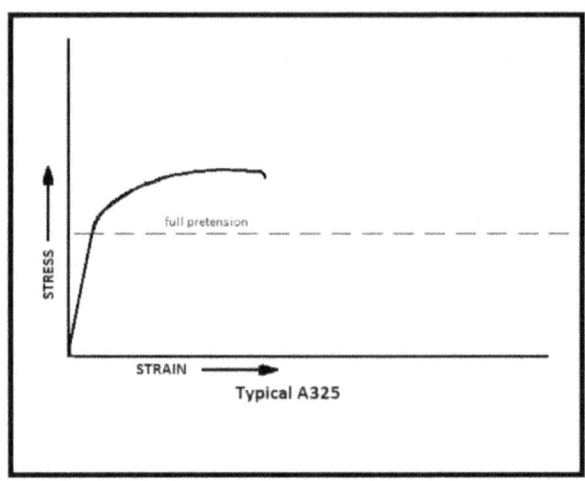

Tolerances?

It is obvious, with all the factors which can affect the actual and final tightness of a joint, perfection will not be likely for every bolt or joint. There are too many things which can make the actual tightness unpredictable to some extent, but a potentially limited extent. Although there can be small unknown factors or effects which cannot all be eliminated at all times, *assuming that one is doing all that they can to eliminate any <u>known</u> problems,* those effects can be limited to a certain margin and controlled to a certain extent. That, of course, NEVER makes it acceptable to do anything which can make those factors worse. Even things which are not perfect may still be "good enough", as long as they do in fact meet the minimum requirements (tolerance). If the minimum requirements are not met, then it is not acceptable. Those requirements were intended to address the potential spread in actual values, and there are factors of safety to account for the <u>unforeseen</u>. But, that assumes that things truly were <u>unforeseen</u>. If not, then the margin can't be assumed to exist, because if one already sees known defects, there are still all the <u>unforeseen</u> ones on top of those. There is a certain required amount of overdesign, such as quantity or size of the bolts, but that is only an overdesign if the potential problems are minimized. If known defects are allowed to exist, then there is no overdesign and the code requirements are not being met. Overdesign is not actually a tolerance, since a certain amount of overdesign is required by code and can't be reduced without creating dangers for the structure. Margins only exist if one makes a <u>full honest effort</u> and there are <u>no known defects</u>. Tolerances are **NEVER** an excuse for not doing things right. Tolerances are only intended to deal with the impossible and unknowable, since all known problems <u>must</u> be corrected. The safety factors only apply after <u>all</u> things were done per procedure and verified to the best of one's knowledge. There is never a tolerance on failure.

There are tolerances for everything in bolting, right down to the chemical makeup of the bolts and the structural steel. There are tolerances for fabrication and erection and fitting. There is what appears to be a tolerance on pretension, but one must be very careful with the interpretation of that, since it applies only to an individual bolt within a joint, and <u>never</u> to the whole joint. Even then, it may be more apparent than real.

Tolerances are meant for factors such as the slight differences in tensions produced in different joints with different grips or different unthreaded lengths. Those factors are generally disregarded, but there is an effect when one rotation is used for a particular bolt that is used in different joints.

Tolerances are meant for matters such as slight deviations of hole alignments which may to some extent be beyond control after all has been done to limit differences. There is 1/16" added to the diameter of the bolt for a "standard" hole, which allows easier erection and accounts for the natural slight inaccuracies of fabrication, but which has only unnoticeable effect from the small loss of material. Plates may not align perfectly, but if they are within a pitch of 1/20 they are considered close enough. That pitch is not considered a danger to the bolt, but there is still an effect.

Because there are tolerances for a number of *individual* things, one must be careful of compounded tolerances which can add up to a problem. Obviously, one should never make such things worse knowingly. Inherent with those tolerances is the assumption that one will do everything as correctly as possible to keep things from getting any worse. If reaming is necessary, which is hard to avoid everywhere but which can be minimized, washers *must* be installed to make up for the loss of material at the hole. That itself could be considered a tolerance in a way, because there may still be effects on the performance of the joint of the now oversized or slotted hole, but it may be considered "good enough," if it is limited, to meet the minimum requirements if the engineer's calculations indicate so. It is clear in this example that one is trying to do what is necessary to resolve the problem and bring it as close as possible to design and not ignoring a known defect.

There is, of course, a huge spread in the actual tightness of individual snugged bolts. That has an effect on the pretension in turn-of-nut pretensioning because each bolt will begin rotation from a different point. It also has an effect, potentially a very serious effect, on the retention of the pretension in previously tightened bolts. If they are all snugged extra tight, then more tension will be retained than if they are snugged less, and they will be far less likely to have tension drop below that assumed in design. When one bolt is tightened, another is being loosened to some extent. If they are snugged a little more, then they will still meet the snugness requirements to a better degree after losing some of that tension as other bolts are tightened, either snugging or pretensioning.

What does a tolerance mean? This obviously refers to only and individual bolt and not an average, which must be right. This does not imply that it is acceptable to have any known noncompliant bolts. Any known noncompliant bolts have a zero percent change of being right. One thing it does mean is that there must be a close control of the spread of tension in a bolt group. It is not allowable to ignore bad bolts just because others are good. Would someone think that it is acceptable and safe to have an entire section of a bridge bad, just because the bridge is so long that one bad section is only a small percentage of the total length? If one were to allow such a situation to exist, it would make no difference how good the other sections are. The whole bridge would be unable to function. Is this example too extreme? Hardly. People actually do think like that and even make such statements that it is just fine to have up to 20% of the welds bad without any repairs necessary.

One can figure that a normal range for relaxation is 3% or more, but it can be made much worse, up to 8% by some studies. 3% or less is what we are trying to get, although it may be difficult to get less than 3% under the usual conditions. After a 5% loss, things begin to get less safe, especially if less reliable methods like "calibrated" wrench are used. When one is looking for the chance that a joint might pass 10% testing, which is the best that can be hoped for with torque oriented methods such as the "calibrated wrench" method, obviously an 8% loss which is common to such methods can be a problem on top of such a large spread of tensions produced, especially when they are only trying to do the least they can get away with.

Snugging becomes a factor during the Mandatory Pre-Installation Verification testing. During the testing the ironworkers always make a greater show of effort than they will do in the actual joints for reasons of the greater fatigue later in the day and greater difficulties up in the air than on the floor. Such a factor can make a more reliable method, such as turn-of-nut with matchmarking, much less reliable. The Pre-Installation Verification can only verify the adequacy of the procedure if it is used exactly, which means using the exact same "full effort" when snugging the actual joints. There is not quite a real definition of "snugtight", but it can be considered to be no less than 10% of full pretension for larger bolts, and much more than that for smaller diameters. When thinking of snugtight as a percentage, while trying to equate it with "full effort" or a few hits with an impact wrench, there will be a difference between an A325 and an A490. What produces 10% of full pretension in an A325 will not do so in an A490. A490 bolts are given a lower reliability rating in the formulas, since they take a greater effort, when people don't even want to use the effort required for an A325, and A490 bolts have a smaller margin.

To account for the much less effort used in actual joints, one should use a lower level of snugness in the load tester during the Pre-Installation Verification to make sure that the procedure that they will actually be using is the one being qualified. (That is for the Pre-Installation Verification only and not the "arbitration" testing, where the 15% value <u>must</u> be used in the load tester.) The contractor should keep that in consideration if they want to joints to pass. There is no tolerance for failure. The joints should be snugged even harder in the actual joints. Also, using a lower snugness in the load tester than they will actually be using in the joints could potentially help to mitigate the effects of relaxation in the snugged bolts as they progress, assuming that the snugging was sufficient to bring the plates into firm contact. But, at the same time, in such a case, one must make sure that they don't get the impression that they don't have to try very hard snugging the joints. If a certain reduction in snugging on a load tester looked good, there is too great a possibility that they won't even use that much effort at the joints. They may misinterpret the point of such an exercise. One must always keep in mind both the purpose and the limitations of testing one bolt at a time in a load tester. One is not trying just to get the needle on the gauge at the right spot on the load tester. One must be thinking about the joints where those bolts will be used. If one only thinks about getting one bolt right, the joint will fail.

In a previous section, there was a "reliability" factor mentioned, which is considered to be 1.13 for all methods used in the calculations, unless qualification tests are done to allow a greater value for the more reliable methods. There is another factor in the equations – the number of bolts in the joint. As either of those increase, then the fraction in the equations decreases and the reliability of the joint increases. Is this a tolerance? Does this allow sloppy work as long as there are more bolts used in a joint? It should be remembered that the result of that part of the equation is always less than one. More bolts simply help the mean bolt pretension in the joint to be closer to the average values determined in the testing.

Let's look at the reliability factor. This is based on conditions which can't always be assumed to exist in every joint on a real project and performed by a typical contractor. If the joint was designed as though the bolts will be slightly tighter than required (5%), then obviously the joints <u>must</u> be tightened above the perceived requirement (by 5% at least), or the joint might not perform as calculated (lower reliability and may not even meet the minimum in reality). At that point there is no margin except positive.

It is assumed (I don't know why since I have never witnessed it) that the contractor will always be *trying* to go beyond the minimum, or it will not meet the minimum.

The idea is based on an *average* value, which must be at least the minimum or greater. The concept of an average is that some bolts may be a little less than others (within limits), but others are more than that so the average meets the minimum. But those which are less than average must still be close to the average determined during the verification testing. Throwing one A490 into a joint of A307 bolts will not be the same as a joint of all A325 bolts. So, it is still necessary that the contractor try to make all bolts meet the requirements, so that if a bolt has a little less pretension than others it won't be by much. Also, if arbitration becomes necessary, it might be the bolt that was tightened less that gets tested (which is the point of that testing), so the contractor should never be relying on the idea that there is a "tolerance" on failure.

If one bolt is looser, others must be tighter by at least that much or more. During the Pre-Installation Verification, one may see how the average is determined and how one must go beyond the minimum to meet the minimum. One will be testing a *set* of bolts, and there will be a spread in values seen on the load tester, either in tension of F1852 bolts, torque values at the required pressure, or angles of rotation (although that is usually the smallest spread). That natural spread is worse for some methods and manufacturers and better for others, and is to some extent beyond one's control as long as all required precautions are followed and one does not do anything to make it worse. All bolts during the Pre-Installation Verification testing, except TC bolts, are turned until right. None, except TC bolts, will be less than required. So then, the average must be above the minimum, and that average torque or rotation qualified is used, and it will be assumed that the pretension will be 1.13 times the required pretension (anything less is not very reliable and will have a much lower chance of actually protecting the joint). If the spread of some methods is smaller, then those methods have a greater reliability.

Apart from the assumption that the contractor will go beyond the minimum a certain amount (at least 5% more and assuming also that they are not using an "effort" control method), it is always in the contractor's interest to *try* and *intend* to go beyond what is believed to be the minimum, and to make the average values determined during the Pre-Installation Verification as far beyond the minimum as practical, or there will be some bolts which fall below the minimum. It is such a theory which places the reliability higher (such as numbers like 1.13 rather than <1), but that only has meaning under the conditions of those tests. It is expected that the inspector will be making sure that the contractor places that average beyond the minimum. It is assumed also that testing will be performed where necessary to be as sure as possible that requirements are being met. Does a number such as 1.13 leave any room for a minus tolerance? Not when that number corresponds to only a 90% reliability rating in other formulas and theories. If understood more correctly, a number like 1.13 is <u>not for the contractor, but the reliability of the verification</u>, since all methods of pretensioning are methods of verification. Some methods of verification, such as matchmarking or DTI washers, are more reliable than snapped splines or a click of a wrench (worst of all). So, it is not the contractor who should be thinking of a reliability rating used in the calculations as a minus tolerance, nor should anyone else. It comes down to an empty theory on the jobsite, since the number does not look as pretty after being run through the formulas (a number like 1.13 times the minimum pretension in one formula may only mean a 90% or worse reliability rating in another formula. The number doesn't look good when done right, and the numbers get quite depressing when done wrong. Those numbers are only part of the equation. The true reliability is on the inspector who must enforce the requirements of the procedure, and the ability of the inspector to find any noncompliant joints before they make the news.

One method of determining the possibility of compliance, when there is a question over whether or not the contractor is following the procedure exactly and effectively or if it is specified for the project, is torque testing, when there was no better method of verification such as matchmarking or DTI washers. This is called by the RCSC specification an "arbitration" torque. It is as unreliable as the "calibrated" wrench pretensioning method, so this method of verification has a lower reliability than I would like.

In this "arbitration" method or testing method, one determines a torque value in a similar way to the Pre-Installation Verification testing. A difference, which the contractors should always keep in mind, is that the arbitration torque is that which is required to rotate the bolt an extra five degrees past the minimum required. This is necessary to help address the unreliability of all torque oriented methods as well as the physics of the process. It was also assumed that the contractor used a target tightness which was 5% greater than the minimum. Therefore, the contractor should tighten the bolts a little more than is believed to be required (at least 5%) or they may not pass. The contractor should be using a torque greater than the arbitration torque, since such an unreliable method can have an excessive relaxation from the huge spread in values in a joint. It is this little bit extra which gives this verification method any value. Therefore, one cannot think of there being any tolerance implied by a number such as 1.13 or by the bolts being tightened slightly more than the perceived minimum. The contractor must try to go beyond the minimum or there will be rejections. Is it fair that the bolts will be tested at a torque which is slightly beyond what produced the required tension on the load tester? Of course it is. It is necessary to compensate for the deficiencies of this method and because it is *expected* that the contractor will be going beyond the perceived minimum by at least that same amount, as is a requirement as noted above, to make sure that all bolts are correctly pretensioned, and the requirements have been based on that assumption. (One can use torque testing after any pretensioning method since it is always possible to establish an equivalent torque. A more reliable method of testing bolts afterward, though, is ultrasonic elongation measurement. Certainty comes at a price.)

One will often see them (sometimes the impression is that it is always, no matter how long they say they have been in this business) simply torqueing bolts in sequence, such as clockwise, rather than using a pattern which minimizes relaxation. One can then test the first bolt and it will fail. When they retorque that one, one can then test the second bolt and it will fail. They may think that it is a coincidence that one can always test the loose bolts first, but it is because they use faulty methods which cause excessive relaxation, which itself is a violation of the provision that their method must minimize relaxation of previously tightened bolts. It takes some planning to get the joints right, such as observing the proper position in the joint to begin with (such the tightest with the least possibility of movement) and then using a cross pattern progressing outward from that point or from the middle when the fitup is more even. Telling them to do it just like tightening a cylinder head or lug nuts on a wheel doesn't help, even though they know what that means. People have a hard time applying the principle when it would slow the process down by a minute or less per joint but save hours after testing. They would rather go back over the joint again after repeated failures, as though they want to see how much sloppy work they can get away with. There is no tolerance for failure, especially when it could almost be considered deliberate. When a method is considered unreliable, one has gone outside the realm of tolerances. Tolerances only relate to a reliability that is greater than necessary on average, or it will extend into the realm of failure.

There is also a misconception with "arbitration" testing regarding the quantity of tests necessary. Testing of 10% of bolts <u>per joint</u> is required, with <u>no less than two per joint</u>. That means that there will be only 10% testing if there are 20 or more bolts in a single joint. Most projects will have much fewer than that, and so there may be as much as 2/3 testing, or even 100% testing if the joints have only two bolts and the contractor can't even do two bolts right. Each joint in question must be tested, so a contractor should never think that there is room for sloppy work. It is better for the project if there is full cooperation with an inspector from the beginning to avoid this as much as possible or delays could result. Someone has to pay for the extra testing which should not have been necessary, and it quickly becomes 100% after the usual number of failures at every joint. This testing can be done on any joint pretensioned by any method if one chooses to establish a torque value on a load tester.

When using turn-of-nut methods, there is what is stated like a tolerance on rotation, such as plus or minus 30 degrees. If 1/3 turn past snug is determined through testing to be what is required, this would be a range of from 90 to 150 degrees of rotation. But, one may say that it is the <u>average</u> we are looking for. That implies that a bolt may have less rotation if another bolt has more. That quickly degenerates to a couple of bolts being pretensioned and other bolts being ignored completely. Obviously there are limits to the range, and obviously, turning a bolt beyond the minimum does not excuse deliberately turning the other less. When looking at that range, people make the potentially fatal mistake of assuming that going more than 30 degrees beyond what is required is dangerous for the bolt and might break it. They try to keep the rotation *below* the required average. The reality is that it doesn't hurt the bolt at all. In times past it had been required to turn bolts far more than 1/3 turn and it didn't hurt the bolts, even if they were turned a full turn. One can look at the mill test reports for the bolts and see that they won't be harmed by an angle of rotation up to several times that required to produce the required pressure with no damage. Danger to the bolts was not the reason for the limits, as is clear from the past examples. The danger, besides the critical danger from bolts not being tight enough, is the potential uneven spread of pressures along the bolts if the range is not kept tight to some degree. If some bolts are too loose and others are too tight, even if the average pressure on the joint (which is the main point of bolting) meets the minimum, the bolts might not share the load evenly enough to work as a team. That can create a danger to the bolts, but nothing like the danger of not meeting the minimum required tension. Is the minus 30 degrees to be considered a tolerance on the required pretension? The minus 30 degree limit still assumes that the minimum tension is there and it comes from the idea that turn-of-nut tensioning is more reliable than other methods and the tension in a bolt will still be close enough, as long as the bolts meet the average together. When performing the Mandatory Pre-Installation Verification it will be observed that there will be a small spread in the rotations of the sample bolts because of various tiny factors. The minus 30 degree limit takes into account the small spread, and the average is assumed by many studies to be more than some of the bolts require, although that is under perfect conditions only and can't be counted on in the real world. Within those limits the bolts can still work as a team. There is no tolerance on failure. Any bolt which is not turned enough can be tested and rejected if it does not meet the minimum. The tolerance is on an individual bolt. There is no tolerance on the average of a joint.

Just because there is a stated upper limit, such as plus 30 degrees, that does not mean that it is rejectable for a bolt to be tighter than required, contrary to popular belief. Overtight is not cause for rejection, and it is safer for a bolt than undertight, as should be obvious from all that has been shown in this work. A couple of bolts overtightened severely may cause load to not be shared equally when other bolts are not tightened as much. That is the real reason for an upper limit, not that the bolt will break or that it is anywhere near the danger of not being tight enough. People often use the excuse of that supposed upper limit to try to stay closer to the lower limit. They forget that undertight is rejectable, not overtight. One should be trying to stay above the minimum as much as practicable to make up for all the other problems which reduce the effectiveness of a joint. As a matter of fact, rotating a bolt beyond the minimum, say 40 degrees, could actually affect the tension less than 30 degrees undertight, because the bolt could cross the yield point where the stress/strain curve becomes nonlinear and it takes far more rotation to equal the same amount of change in tension. Overrotation has far less of an effect than underrotation, and one should be concerned for the bolts not being tight enough. Too tight is not cause for rejection.

As stated previously, just because 1/3 turn can produce 51 kips of tension in a particular bolt, or pressure on a plate, that does not mean that 2/3 turn would produce anything near 102 kips (the increase in tension is quite small). It won't affect the tension much. But if 2/3 turn is required as demonstrated in the Mandatory Pre-Installation Verification, then only going 1/3 of a turn will lower the tension a huge amount because it is in the elastic range where a small change in elongation has a huge effect as seen by the steep slope of that part of the curve.

One should be trying to keep the spread small so that there is an even distribution of stress throughout the joint. Having a bolt or two too tight does not make up for loose bolts. Using an A490 does not mean that the other bolts can be A307s. Bolts which are too tight will take a greater proportion of the load while other bolts take less. That would be dangerous if is too extreme. Also, making some bolts too much tighter will take tension off of other bolts. An extremely overtight bolt will desnug the other bolts and the angle of rotation will be meaningless. So, there is a limit of a 60 degree spread of rotation for shorter bolts, and other limits for longer bolts, to the spread of values and keep the tension in the bolts more even. There are cases where the average doesn't help the joint, since an average in extreme cases may not give a true picture of the real situation. Too big of a spread is dangerous no matter what the average is, since too big of a spread may mean that the joint can't take a load as designed.

 A related matter is the use of DTI washers, where there are to be a certain number of bumps crushed, but it is said that the contractor should be careful not to completely crush all the bumps. One could possibly view this as a tolerance in a way, since it gives a spread between most and all. But, there is only a positive tolerance implied, since the bolts must still all meet the minimum pressure. More is still much better than less in this case, and too much is still not cause for rejection. If one does not go far enough, the tension can't be measured or even assumed. If one goes a little further than the minimum, one does know that the minimum was reached. Even crushing all the bumps down won't hurt the bolt, as long as someone does not go too much farther into that positive range. Of course, there is a risk of a bolt carrying too much tension to share the load with other bolts not turned as far, and there could be larger spread in actual pressure across the joint if the bolt rotation is not kept within a narrower range.

 There is, of course, a small spread in the values of pressure required to crush DTI's, and the mill cert will show that the average value is above the minimum required to be certain that it won't be less. Does that mean that one is allowed to be sloppy and that less is tolerated? No. Among other things, one never knows if an individual DTI washer will require an excessive pressure or not. One should, as always, be trying to go beyond the minimum by at least 5%, that being the actual minimum, if not more, just to fight relaxation and to be more certain that the bolts will all do their job. Even if the average pressure appears to be right, a bolt which is too loose will contribute nothing to the load capacity of the joint until the other bolts have already yielded or even broken, especially the tightest ones, and so there can be no tolerance on anything less than the minimum of the acceptable spread, no matter what the average is. It should be clear that the average must be above the minimum if all the bolts are to meet the minimum and share the load. One can never count on there being anything extra in the end, even if a few bolts, if any, are tighter than required. If it wasn't for that little extra, few bolts would meet the minimum requirements when the contractor uses the standard principle of doing no more than what is believed to be the minimum but never actually is. It is partly because of the average pressure of DTI washers being designed above the minimum (actually only a couple kips) that this method has a high *potential* reliability. Even if the average value for DTI washers, according to the mill cert, is more than the minimum, one cannot assume that the bolts will all be above the minimum by that amount. We are still talking about an average of a spread of values, and that average includes DTI washers which crush at a lower pressure than others. The average must be a little more than the minimum or even the DTI washers won't meet the minimum, no matter what the average appears to imply. The extra is only an illusion.

Can one go too far with DTI washers? It is perhaps more likely than turn-of-nut. Except for the paint squirting type, one has to turn the nut and then check the washer, and one does not know the moment when the minimum level is crossed. It is often only a guess, if one isn't paying attention to rotation which someone should have been measuring during the Mandatory Pre-Installation Verification. When that level is crossed, one only knows that it was crossed and not what the actual pressure is. One only knows that it is above the minimum. One must be trying to do more than the minimum unless one doesn't mind going back to the same joint after repeated failures. This is a part of why DTI washers can be considered more reliable than turn-of-nut, with another part being that it measures actual pressure created by the bolt when other methods are only a guess. It can also measure pressure *through* the joint, since they are placed under that part not turned. Also, all bolts must still pass the minimum, and one can be much more certain that the bolts are creating the necessary pressure.

As with all things in bolting, making the bolts too tight is not a bad thing in itself. Once the bumps are all crushed the pressure can be anything. Overtightness is not the real concern but the possibility of an imbalance in the joint if the spread of values becomes too large. If all bolts are tightened to the same degree and all are above the minimum then the risk to the joint is much lower than if the bolts were not tight enough. There is a risk of imbalance with this method (usually only on the positive side which is not a problem and not very significant with most contractors) but the risk is probably smaller with this method than with others.

With turn-of-nut, the amount of pressure produced by overrotation may be small. With DTI washers, because it is the pressure being tested, a higher than required pressure will correspond to a much larger rotation according to the stress/strain curve. It is not so easy to go too far with this method, especially with larger diameters.

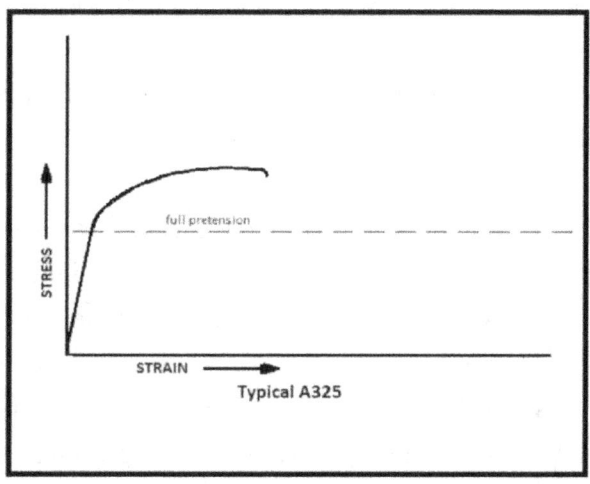
Typical A325

Concluding Remarks

The codes were written in blood, and one must never forget that. If that is ever forgotten, then all we have gained in knowledge from past mistakes won't help anyone avoid those mistakes.

Bolting is very simple. All one has to do is: 1) perform the necessary Pre-Installation Verification, 2) bring the members together, 3) line up the holes, 4) insert the bolt, and 5) turn the nut. People make it harder than it needs to be. If there is a failure, then it relates to one of those five steps, assuming that one is using good bolts and the members were fabricated properly. The most common failure is at the turning stage, and it is often even deliberate.

One does not have to understand much to do it right. All one has to do is follow the procedure. It is that simple. But, it is that lack of understanding among contractors, inspectors and sometimes engineers which is the reason why procedures are disregarded. And, if inspectors learned to understand these principles, then the contractors would also, or they would need to learn before their budget runs out.

The engineers shouldn't be required to understand much of the procedures in bolting in fine detail, as though they had worked as ironworkers themselves, but it does help. The engineers should only have to worry about the design of the joints and such things as loads, strengths or configuration of the steel. It is up to the contractors and inspectors to understand how to make it happen. Accurate calculation and top-of-the-field design methods have no meaning if the design is not followed in all of its principles. A building or other structure cannot ever be any more functional or reliable than the contractor makes it, and as the inspector verifies it.

If some of what I described regarding the real world seems extreme, then too bad. I didn't exaggerate. Are things really that bad? It is inexcusable that they are. If you don't like it then fix the problem.

The matters brought up in this guide are not intended to teach how to design a joint. The purpose of this is not to add another design manual to the ones out there already. I am only trying to make people aware that all things matter, and all matters, no matter how tiny and seemingly trivial, which people are oblivious to, make a difference, either positive or negative. Many of these things are not considered on a jobsite, but it is important to know these things about design to better apply the intentions of the design. So much of the difference factors make is outside of control in production and erection tolerances that what can be controlled is even more critical if there is to be any remaining margin. Everything matters, down to the things no one thinks about. It is all parts working together to make the whole what it is and how it behaves.

Shoving a bolt in a hole isn't the beginning. Snapping a bolt isn't the end of what is taking place. The bolt is only a piece of a system. The bolt itself isn't even the real point. Having that system either perform as designed or fail with unpredictable costs is what we are talking about.

FURTHER READING

This was an experience based work, but there are many resources available for those who want to delve deeper into the design of high strength bolt connections and behavior. Besides the excellent commentaries in the RCSC Specification and others such as the AISC Seismic Provisions, there are countless articles and books, of which I felt that a few would be helpful for those trying to see bolting in a clearer light. I will list a few below.

One can hardly do better than to read from the masters of bolting who wrote the foundation of current design, even if their conclusions give a far rosier impression than experience shows – Geoffrey L. Kulak, John W. Fisher, John H. A. Struik, "Guide to Design Criteria for Bolted and Riveted Joints", 2nd edition, John Wiley and Sons, New York, NY 1987.

See also:

Weiyan Tan, Vladimir V Maleev, Peter C Birkemoe, "Installation Characteristics of ASTM F1852 Twist-off Type Tension Control Structural Bolt/Nut/Washer Assemblies," Final Report phase I, Department of Civil Engineering, University of Toronto, Toronto, ON 2005.

Schmeckpeper, Edwin R. et al. "The Effects of Over-compressing ASTM F959 Direct Tension Indicators on A325 Bolts used in Shear Connections", Engineering Journal, first quarter 1999.

Wayne Wallace, J and M Turner, Inc, "Torque Won't Discover Loose Bolts" American Consulting Engineers Council's Heavy Movable Structures Movable Bridges Affiliate 3rd Biennial Symposium November 12th-15th, 1990.

Wade, Patrick Michael, "Characterization of High-Strength Bolt Behavior in Bolted Moment Connections" (under the direction of Emmet Sumner) North Caroline University, Raleigh, 2006.

Silviu-Christian Melencius, Vasile-Mircea Venghiac, Andrei-Ionut

Stefancu, Mihai Budescu, "Factors Influencing the Preload Level of High Strength Bolts for Structural Steel Connections" Technical University of Iasi, 2011.

www.ingramcontent.com/pod-product-compliance
Lightning Source LLC
Chambersburg PA
CBHW062216220526
45471CB00009B/3227